Mid-century Plastic Jewelry

Susan Maxine Klein

Principal Photography by Jori Klein

Schiffer Publishing Ltd

4880 Lower Valley Road, Atglen, PA 19310 USA

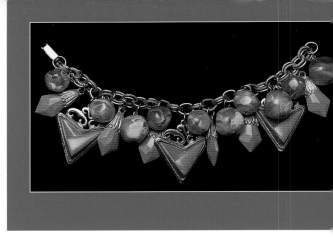

Library of Congress Cataloging-in-Publication Data

Klein, Susan Maxine.
  Mid-century plastic jewelry / Susan Maxine Klein ; principal photography by Jori Klein.
    p. cm.
  ISBN 0-7643-2234-6 (softcover)
1. Plastic jewelry—Collectors and collecting—United States. 2. Costume jewelry—Collectors and collecting—United States. I. Klein, Jori. II. Title.

NK4890.P55K57 2005
688'.2'0973075—dc22

2005007258

Designed by Mark David Bowyer
Type set in Bauhaus Hv BT / Humanist521 BT

ISBN: 0-7643-2234-6
Printed in China
1 2 3 4

Published by Schiffer Publishing Ltd.
4880 Lower Valley Road
Atglen, PA 19310
Phone: (610) 593-1777; Fax: (610) 593-2002
E-mail: Info@schifferbooks.com

For the largest selection of fine reference books on this and related subjects, please visit our web site at **www.schifferbooks.com**
We are always looking for people to write books on new and related subjects. If you have an idea for a book please contact us at the above address.

This book may be purchased from the publisher.
Include $3.95 for shipping.
Please try your bookstore first.
You may write for a free catalog.

In Europe, Schiffer books are distributed by
Bushwood Books
6 Marksbury Ave.
Kew Gardens
Surrey TW9 4JF England
Phone: 44 (0) 20 8392-8585; Fax: 44 (0) 20 8392-9876
E-mail: info@bushwoodbooks.co.uk
Free postage in the U.K., Europe; air mail at cost.

Dedication

This book is dedicated first and foremost to my late mother, Leah Moltz Klein, who inspired all this wonderful madness. She was an antique hunter extraordinaire; I learned from the best. My father, Dr. Samuel Klein, and my brother, Dr. Frederick Klein, have also been inspirational in my taking a family hobby of antique collecting to an exciting career.

This book is also dedicated to the costume jewelry workers of Providence, Rhode Island, for their craftsmanship in creating such objects of incredible beauty.

Disclaimer and Acknowledgments of Trademarks

Most of the items and products in this book may be covered by various copyrights, trademarks, and logotypes. Their use herein is for identification purposes only. All rights are reserved by their respective owners.

The text and products pictured in this book are from the collection of the author of this book, its publisher, or various private collectors. This book is not sponsored, endorsed or otherwise affiliated with any of the companies whose products are represented herein. They include Coro, Trifari, Lisner, Kramer, Selro, Marshall Field's, among others. This book is derived from the author's independent research.

Since many post-war plastics can be hybrids or blends (i.e. Lucite® combined with styrene) and others would be difficult to ascertain without using invasive, destructive, and costly testing methods, the specific composition of the plastic jewelry will only be mentioned when identification is assured through appropriate testing.

Values, unless otherwise noted, are for costume jewelry in very good to excellent condition. Values given are of a retail, national average; however, valuations in major metropolitan areas (particularly on the U.S. east and west coasts) can be slightly higher.

All jewelry is from the author's collection, unless otherwise indicated. Photography by Jori Klein, unless otherwise noted.

The author's antique-loving mother Leah Moltz Klein. All photos, unless otherwise indicated, have been taken by Jori Klein.

Acknowledgments

So many people have helped bring this idea from proposal to published tome. My sincere appreciation to all who helped me collect, tell the story, and bring this project to fruition.

My family played a major role in the entire publishing process. An extra special thanks goes to my brother, Frederick Klein, who has been my chief collector and provided both emotional and financial support to this project. You are my favorite co-conspirator in antique hunting! A second extra special thanks goes to my niece, Jori Klein, for providing such magnificent principal photography. You truly brought the subject matter to life! I would also like to thank my sister-in-law, Sharon Klein, and niece, April Specht, for providing editing and proof reading expertise. Last but not least, my parents, Samuel and Billie Klein, for letting me bore you with every detail of the process.

Next, I must thank friends - Karen Prena for being my brilliant legal counsel and idea-person extraordinaire, and Jane Endres, Shawn Marsh, and Lisa Wise for sharing my antiquing habits. I thank my former boss at Sotheby's, Larry Sirolli, for kindly writing the Foreword, and my current bosses, Richard P. Norton of Richard Norton Gallery and Richard M. Norton of Richard Norton, Inc., for being so supportive in this endeavor. A special thanks goes to Leslie Hindman for all her encouragement.

Thanks also go to the lovely ladies of the Chicago Vintage Fashion & Costume Jewelry club (VFCJ): Cheryl Killmer, Judy Levin, and Carol Sullivan for sharing their fabulous collections; Wendi Mancini for sharing her historical Trifari information; and Ellen Germanos, for getting me involved in such a wonderful organization. Additional thanks go to Sheldon Atovsky for allowing us to photograph magnificent pieces from his store, Studio V; to Francesca Medrano for providing pictures of her outstanding collection; and to Pat Seal, of Treasures From Yesterday, for sharing her vintage ads and vast knowledge.

The next group of people I thank helped me tell this story. Your knowledge and memories brought the subject to life, and for this I am deeply grateful. Many thanks go to Tony Jahn of Marshall Field's in Chicago for allowing me to access your incredible archives, and to my principal plastics expert, Maureen Reitman, Senior Managing Engineer of Exponent Inc., for helping me to understand the properties and possibilities of plastics. Very special appreciation goes to Jane Civins of The Providence Jewelry Museum for sending me numerous publications, articles, and wonderful archival materials. The following people have been invaluable to this project: Robert Andreoli, John Asencio, James Axelrad, Al Bagdade, Glenn Beall of Glenn Beall Plastics, Kate Ganz Belin, Marsha Brenner, Milton Brier, Linda Cerce, Steve Coulter of Lucite International, Bob Dubinsky of Smart Creations, Victoria Ganz de Felice, Frank DeLizza, Karl Eisenberg of Eisenberg Ice, Jack Feibelman, Tony Ganz, Iraida Garey, Bob Gluck of Perfect Pearl, Burleigh Greenberg, George Ivanov of Panduit Corp., Larry Kassoff, Jim Katz of Katz Jewelers, Norm Kaufman, Bane Kesic of Molex Corp., Geraldine King, Dennis Knight of the Providence Jewelry Museum, Keith Lauer of the National Plastics Center and Museum, Sally Loeb, Steve Lord of Greene Plastics, Robert Mandle, Alan Marcher, Joan Marcher, Jean Burritt Robertson, Director of Research and Development for the Rhode Island Economic Development Corp., Rodney Rouleau of Panduit Corp., Carol Rowe of Plastics from the Past, Carole and Stan Smith of Ralph Singer Jewelry, Ed Sternberg, Joan Talmy, Lucille Tempesta of VFCJ, Ron Verri of GemCraft, Irving Wolf, and Nancy Ganz Wright. I thank you all for speaking with me about this exciting subject!

Contents

foreword

Gregarious, willing to lend a hand, enthusiastic and passionate – traits that immediately come to mind as I ponder Sue Klein and this Foreword. Sue is a professional colleague and a friend. Her book on plastic jewelry produced in the 50's and 60's is not as odd a collecting category as you might think for a senior decorative arts specialist and 22 year veteran of Sotheby's to be introducing. We are fellow collectors; we share the same appreciations and engage in similar behaviors as collectors.

My first auction purchase and foray into collecting was a 75 pound snapping turtle, which friends and I purchased in Philadelphia at the age of 14. Fifteen years later, prior to seeking a career in the auction business, I spontaneously purchased an Art Deco vase with savings set aside for a bicycle. I have moved on from turtles, but still collect rocks and other works of art.

I met Sue upon arrival in Chicago as I took the helm of Managing Director of Sotheby's Chicago. The staff, a combination of familiar and unfamiliar, seasoned and freshly hired specialists and administrators excitedly welcomed me as part of their group. Sue introduced herself and proceeded to engage me in conversations about art, collecting, and her passion for industrial-made plastic jewelry. She shared a dream to author if not the first, an informative book on this little known, inexpensive, sought after jewelry. She was wearing her latest acquisition.

All collectors and scholars, regardless of their field, be they collectors of French furniture, Impressionist paintings, baseball cards, or hat pins must be passionate, knowledgeable and willing to share information. Sue possesses enthusiasm and eagerly shares her knowledge of the jewelry she so passionately collects.

It is with great pleasure and an honor for me to introduce to you Sue's eagerly awaited, thoroughly researched book.

Enjoy—

Larry J. Sirolli

Senior Vice President and Director, Sotheby's Arcade

Introduction

I have been attending antique shows since I could first walk! My mother was a doll collector and my father and brother enjoyed military memorabilia. The other collections that filled our home included samovars, Indian pots, antiquities, and art glass. When I was in grade school, my mother would hand me a $5 dollar bill at an antique show to spend on whatever I desired. She taught me how to bargain for and spot the best deals. She was, however, very unhappy with me when I once spent the entire $5 on a "Gone With the Wind" movie program; she could not believe that I spent $5 on a 'piece of paper!' Another of my $5 purchases was a reverse carved, apple juice Bakelite dress clip! I guess I was intrigued by costume jewelry even back then. As I grew up, however, I viewed antiques as those things that cluttered up the house. I no longer wanted to be part of the family antiques collecting madness. I graduated from college with a journalism degree and for about ten years collected nothing at all.

One day, the collecting bug bit me. I was working for the Public Broadcasting Station in Chicago, which is located in a quiet, residential neighborhood on the far north side of the city. There were few stores to visit on my lunch break, but there was a thrift shop. I wandered in, since there were no other places to peruse. Before long, I was spending my entire lunch hour there. It was such a thrill to spend a few dollars on a treasure! I liked collecting depression glass, old cookbooks, and small, vintage kitchen appliances. My home was beginning to fill with treasures!

Finally, I made a conscious decision that if I brought something home, I would have to actually use it. One day, while attending a flea market, I spotted a peculiar plastic bracelet. It had chunky silver links topped with light blue, plastic half-moons. It cost only $10 and it was mine. Mid-century, plastic jewelry was plentiful and affordable; soon a major collection was born. I viewed costume jewelry as the perfect collection: I could actually wear it and yet the pieces could be tucked away in a drawer, out of sight. But 600 plus pieces later, you can't tuck it away! This grand obsession eventually led to my work with Leslie Hindman Auctioneers, Sotheby's Chicago, and as a production assistant on the HGTV Program "The Appraisal Fair." My Sotheby's colleagues were amused by my funky, plastic bracelets. For a time, I shared an office with the jewelry department. I knew my plastic jewelry was something special when the women in the jewelry department, who handled the real stuff all day, demanded to try on my jewelry.

As I searched for information about modern plastics, I found that little was available. Though there were books about Celluloid® and Bakelite®, there were none that exclusively featured the mid-century thermoplastics and thermoset plastics. I have finally put that journalism degree to good use in telling the true story of mid-century, plastic jewelry. Like streamlined cars of the era, costume jewelry reflects a prosperous time: it is bold and flashy. When the Depression and World War II finally were over, the jewelry industry in America exemplified the quintessential American dream. This is a story of industrial America and the ingenuity of the men and women, mostly immigrants, who hand-manufactured fine products. We must, however, start with the Pilgrims, for the story really begins with them.

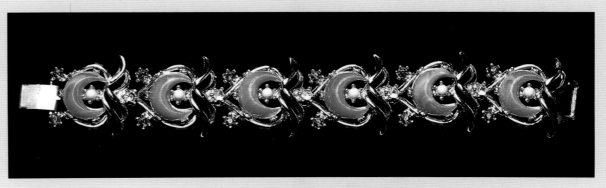

The bracelet that started it all. Link bracelet, featuring light blue, plastic half-moons. Unsigned. c.1950s. $35-40.

Determining Values

Condition is probably the most important factor in determining valuation. The highest range of value is given to a piece in perfect, unworn condition. An original box or hang-tag adds to the estimate of a piece of costume jewelry. Obvious wear, darkened glue or discoloration on plastic parts, and missing plastics parts, stones, or findings contribute to diminished worth. Properly replaced missing stones or plastic parts generally do not diminish value. Rarity is also a factor in determining worth. An uncommon color of a fairly common piece can greatly increase the valuation. Finally, provenance, when absolutely determined, can boost the value of even the most common of pieces. For example, a piece of costume jewelry that was purchased from the estate of a famous person (i.e. you have both the bill of sale and a picture of the celebrity wearing the piece of costume jewelry) would add great value to a rather average piece.

Your Treasure Hunt

You should start your search for costume jewelry at home. Your mother, great aunt, or grandmother's jewelry boxes are the best place to start (with their permission, of course). Thrift stores, garage sales, rummage sales, estate sales, antique stores, and on-line and live auctions are excellent sources for building your costume jewelry habit. Remember, a collection starts with one intriguing piece and you, too, will be bitten. Happy hunting!

1.
The Costume Jewelry Industry of Providence, Rhode Island

Pilgrim emigrants of the seventeenth century, escaping religious persecution in England, brought to America knowledge of making silver buckles for shoes and hats. In 1775, silver items were produced by colonists in the Providence, Rhode Island, area.[1,2] The seaport of Providence was bustling with trade, and by the late 1700s Providence silversmiths were crafting sailor's booty of gold and silver coinage into useful and identifiable items.[3]

"These early capitalists were afraid that if their hordes of coins were stolen they could not easily identify their own shillings or pieces of eight. Therefore, they had their coins fashioned into engraved silver plate such as teapots, trays, tankards, and other items. Engraving the pieces allowed them to be much more easily identified if stolen. This practice encouraged many skilled silversmiths to settle in our colony, which probably set the stage for the development of the jewelry industry here."[4]

In 1793, half brothers Nehemiah and Seril Dodge opened Providence's first jewelry making shop and subsequently developed a method for plating base metals with precious metals. A less expensive metal, like copper, could then be plated with gold or a gold alloy. In 1798, Nehemiah Dodge began selling this cheaper "plated" jewelry, thus officially beginning the costume jewelry trade in Providence.[5,6] By 1947, 90% of the costume jewelry produced in the United States came from Providence and the surrounding area.[7] It is of historical note that the Dodge brothers' apprentice was Jabez Gorham, who founded the Gorham Silver Company in 1831.[8]

Plastics Used in Jewelry

"I just want to say one word to you. Just one word: plastics."

This infamous line from the movie *The Graduate* exemplified a social revolution within America in the 1960s; however, it was the industrial revolution of 1860s America that allowed the Providence, Rhode Island, costume jewelry industry to eventually adopt the use of plastics. Although this book features plastics of the mid-20th century, it is important to understand how the plastics industry began in the mid-nineteenth and then developed in the twentieth centuries.

Parkesine and Celluloid: The First Semi-Synthetic Plastics

In 1845, German chemist Christian Schönbein laid the basis for what would become the modern plastics industry. Due to a laboratory mishap, Schönbein wiped-up some spilled chemicals with a cotton apron. When he left it outside to dry, the apron exploded.[9] Schönbein's experimentation with cellulose nitrate (gun cotton) for use as explosives led Londoner Alexander Parkes, in 1862, to further develop cellulose nitrate for other uses. Parkes combined cellulose nitrate with oils and solvents with the intention of producing novelty items and subsequently developed the first semi-synthetic plastic: Parkesine. Parkes introduced his cellulose nitrate-based combs, buttons, and boxes of Parkesine at the 1862 International Exhibition in London. Although Parkes received a patent for his work in 1865, Parkesine was a commercial failure. His use of poor quality raw materials led to an inferior product that warped and cracked. Parkes was out of business by 1868.[10,11]

The first commercially successful, semi-synthetic plastic was Celluloid®. In 1863, billiards supplier Phelan and Collander offered a $10,000 prize for the development of synthetic ivory billiard balls, since ivory was becoming scarce and expensive. Journeyman printer John Wesley Hyatt, Jr., of Albany, New York took the challenge. Although it is unclear if Hyatt ever claimed the prize, he filed a patent for an invention on October 10, 1865 and formed the Hyatt Billiard Ball Company. In perfecting his billiard ball, on June 15, 1869, Hyatt patented the material we know as Celluloid, a mixture of collodion (cellulose nitrate and ethyl alcohol) and camphor that was heated to form a solid. Hyatt's invention was extremely successful and soon the Celluloid Manufacturing Company was making numerous commercial products. By the 1870s, dentures, dresser sets, eye glasses, shirt collars, buttons, hair combs and, of course, costume jewelry were among the items being made of Celluloid. According to J. Harry DuBois,

Amber and ivory Celluloid-type snake bangles, adorned with rhinestones. French. Unsigned. c. 1920s. Jewelry courtesy of Studio V, Chicago. Amber snake $125-150. Ivory snake $200-250.

in *Plastics History U.S.A.*, an item of costume jewelry called 'South American Jewelry' was Hyatt's first profitable venture. Celluloid, it should be noted, was the trade name of the Celluloid Manufacturing Company. Much beautiful jewelry came from the Celluloid Company itself and other makers of Celluloid-type (cellulose nitrate or pyroxylin) plastics. Perhaps, because of its great success in the marketplace, no matter the manufacturer, jewelry made of cellulose nitrate-type plastic material is commonly called Celluloid today. It can often be distinguished by a camphor-like smell. U.S. production in this medium extended into the late 1930s; however, Japanese manufacturers produced Celluloid jewelry until the 1950s. Celluloid successfully copied ivory, jet, tortoise shell, coral, and amber. Thus, these plastic consumer products were made affordable for everyone.[12,13]

Celluloid, being highly flammable and explosive, was not suited to the growing need for substances that could withstand heat. Although a non-flammable version of Celluloid, cellulose acetate, was made commercially available in 1927, it did not have heat-resistant and insulating properties.[14]

Bakelite: The First Fully Synthetic Plastic

The advent of electric lights and automobiles made Bakelite® a commercial and household necessity. Leo Hendrick Baekeland was a Belgian scientist who sought to develop a substitute for shellac. He discovered a substance, according to *Time Magazine's* "The Time 100: Scientists and Thinkers," that "transformed the world." Bakelite was Baekeland's second incredible invention. In 1899, for the astonishing sum of $1 million dollars, George Eastman, of Eastman Kodak, bought the rights to Baekeland's Velox photographic paper, which enabled pho-

tographic prints to be developed without the use of sunlight. Baekeland then built a house and a laboratory in Yonkers, New York, and set out to develop synthetic shellac. Imported from Southern Asia, shellac, a natural substance deposited on trees by the secretions of beetles, was becoming expensive and scarce. Shellac proved to be an effective insulator for the new and burgeoning electrical industry, with more demand than supply. Baekeland was able to control the condensation reaction between phenol and formaldehyde in a pressure cooker-type device that he called a "Bakelizer". Upon initial heating, phenol and formaldehyde produced a shellac-like substance that, when placed in the "Bakelizer" and further heated, became hard and moldable. This proved to be a perfect substance for electrical components and automobile parts. Baekeland patented his invention in February 1909, and later that year, announced his discovery at a meeting of the New York chapter of the American Chemical Society.[15] Bakelite®, a phenolic resin, became widely used in jewelry production until the early 1940s, when the raw materials became unavailable due to military needs of the Second World War.[16]

Like Celluloid, Bakelite is a trade name and a generic description for plastic as well. Although there were numerous makers of cast phenol-form-

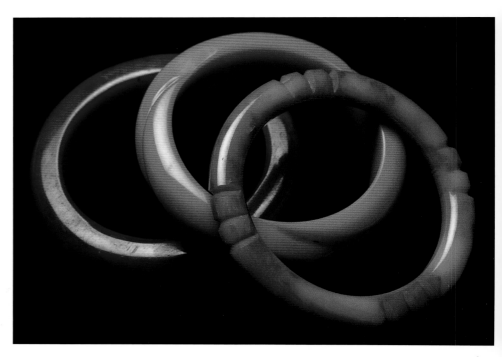

Three Bakelite-type bangles in red, orange and green. c.1930s. Maker unknown. $50-75, each.

Asian-themed, Bakelite-type charm bracelet. Unsigned. Probably c. late 1930s-early1940s. Although the look of this bracelet speaks of this timeframe, it could well have been assembled post World War II, of old, existing parts. During and just after the war, jewelry findings and rhinestones became scarce due to the lack of base metals available for manufacture and the war's disruption of the Czechoslovakian rhinestone market. $45-55.

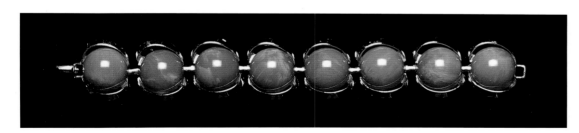

Bracelet of green globes, featuring Bakelite-type material. Unsigned. c. 1940s. $45-55.

aldehyde jewelry (i.e. Marbelette Corp., Catalin Corp.), it is commonly known as Bakelite. Bakelite can often be distinguished by a formaldehyde smell. Leo Baekeland's son, George, sold the Bakelite Corporation to Union Carbide in 1939.[17] Original Bakelite colors were dark; it was not until 1928, when the American Cyanamid Company introduced "Beetle" urea-formaldehyde resins, that brightly colored Bakelite-type material was readily available in the United States. Most urea plastics can be characterized by a fishy smell; however, the earliest versions of urea plastics (thioureas) have a sulphur smell.[18, 19]

Casein Plastics

A close cousin to phenol-formaldehyde plastics are casein plastics. These are semi-synthetic plastics created with formaldehyde and milk protein. In 1897, German scientist Dr. Adolph Spitteler, in an effort to make a white blackboard to replace black school slates, developed casein plastics. He ended up with a product that became a substitute for horn, and buttons and buckles were then commonly made of casein. Under the trade name Galalith®, much spectacular jewelry in the Art Deco style was produced. Galalith has become the name of common usage for casein-type plastics. Casein plastics, not surprisingly, often have a smell like burned milk. Casein type jewelry was widely manufactured in Europe throughout the 1920s and 1930s.[20, 21]

Cellulose Acetate

In the early 1900s, Swiss brothers Drs. Camille and Henri Dreyfus did experiments toward developing a non-flammable photographic film, which could replace highly flammable cellulose nitrate film. In 1910, they produced cellulose acetate film and were contracted to develop this non-flammable film stock for the French motion picture company Pathé Freres.[22] At the same time, the brothers were also using cellulose acetate to stiffen airplane and zeppelin fabrics.[23] During the First World War, their company grew quickly and eventually became the companies British Celanese Ltd. and the Celanese Corporation of America. In 1924, they developed a cellulose acetate fiber, Celanese®, and in 1927 a cellulose acetate molding compound trade named Lumarith®. This substance eventually replaced dangerous cellulose nitrate in toys, jewelry, and other household goods.

After John Wesley Hyatt, Jr.'s death in 1920, Celanese Corp. acquired the controlling stock in the Celluloid Mfg. Co., and in 1927, they bought out the remaining shares. Cellulose acetate was the first plastic to be injection molded. It is available in a wide range of colors, can be characterized by a vinegar smell, and is actively used today in jewelry and other plastic materials production.[24, 25] Both earlier (1960s) and more contemporary examples of jewelry produced by Parisian designer Lea Stein are made of laminated cellulose acetate.[26]

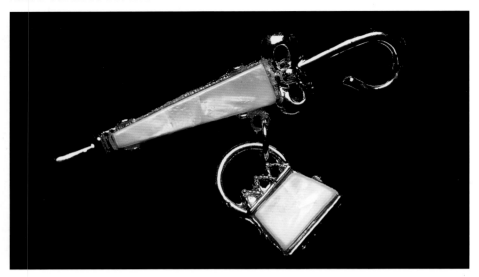

Cellulose acetate umbrella and handbag pin. Upon initial examination, the material appeared to be of a Celluloid-type (Cellulose Nitrate or Pyroxylin plastic); however, when it was tested in the FTIR spectrometer at Analyze Inc. (See End Note #54), it tested conclusively as cellulose acetate. Unsigned. c. 1950s. $30-35.

Beige fox pin of laminated cellulose-acetate. Lea Stein.c.1970s. Jewelry courtesy of Judith Levin. $50-75.

White, feline pin of laminated cellulose-acetate. Lea Stein. c. 1970s. $50-75.

Springy, Cellulose acetate snake bracelets in orange and blue. French. Unsigned. ca. 1930s. Jewelry courtesy of Studio V, Chicago. $50-75, each.

The Acrylics: Lucite and Plexiglas

Acrylic plastics (the most common of which is polymethyl methacrylate) were originally developed in Germany by Dr. Otto Röhm in 1901. His company, Röhm & Haas, introduced Plexiglas® in 1931. The American company E.I. DuPont de Nemours & Co. (DuPont) of Wilmongton, Delaware, entered the field of acrylics with their brand, Lucite®, in 1937.[27] While DuPont used the Lucite trade name in 1926 for a line of Celluloid-type (cellulose nitrate) toilet articles, Lucite, the acrylic, was originally called Pontalite.[28] Acrylics, derived from petroleum products, make an excellent glass substitute and were used extensively for this purpose.[29] Today, acrylic jewelry is commonly referred to as Lucite. Acrylics are characterized by having no smell and are generally colorfast. A piece of transparent Lucite-type jewelry from the 1930s remains clear and not yellowed nearly 70 years later. Acrylics come in a wide range of colors and can be opaque as well as clear. According to Steve Coulter, representative of Lucite International, the current owners and manufacturers of the Lucite® product, "Acrylic doesn't age as badly as other plastics."[30]

Lucite-type bangles, earrings, and pin. The author's grandmother purchased the pin for her at Woolworth's. All unsigned. All c. early 1970s. Smooth bangle $35-40. Carved bangle $40-50. Earrings $15-20. Pin $25-30.

Carved, Lucite-type swimming fish pin. Unsigned, possibly Little Nemo. Jewelry courtesy of Studio V, Chicago. c. 1940s. $120-125.

Polystyrene

Polystyrenes, like the acrylics, are petrochemical-based plastics. Styrene, itself, is a natural substance derived from distilled Styrax tree resin. In 1839, Eduard Simon, a German apothecary, discovered polystyrene, which he called Styrol.[31] In 1869, French scientist Marcellin Berthellot first synthesized the substance. In preparation for war, as a component of a substitute for rubber, styrene production was begun in Germany by I.G. Farben (BASF) in 1929. Dow Chemical brought polystyrene to the U.S. in 1937. Polystyrene could be injection molded and became very popular in the production of mass-produced plastic products, such as inexpensive house-wares, toys, and jewelry.[32] Polystyrene is characterized by its rigid, brittle nature and does not have a discernable smell. It is available in a full range of colors. When molded, polystyrene can exhibit a severe parting line; this is the place where the two halves of the mold meet.[33]

Epoxies

Epoxy resins are a rather late development in plastics production. They were invented and patented in 1939 by a Swiss chemist and dentist, Dr. Pierre Castan, for Ciba Ltd. The resins were introduced to the marketplace in 1946. Epoxies are commonly used as glues because of the excellent adhesive properties. Epoxy-based compounds were developed for dental fix-tures and castings, but it found multiple useful applications for surface coatings, automobile components, and glue itself. [34,35] In the 1970s, jewelry designers began to use epoxies more readily in jewelry production.[36] Epoxies can be tinted any color, although they generally have a yellowish tinge. As epoxies age, they become even more yellow. Epoxies have no significant smell.[37]

Other Things To Know About Plastic Jewelry
Besides the tell-tale look and smell of certain plastics, addi-tional physical properties can further aid in their identification. They are classified as either thermoset or thermoplastic. The plastic jewelry featured in this book, c. 1948-1970, utilized both thermoset plastics and thermoplastics.[38] Identification, however, can be difficult and often cannot be absolutely achieved with-out using destructive or expensive testing methods. Although we will further discuss testing methods, it is best to just enjoy your jewelry and know that they are all "thermo's" of some sort.

Thermoset
Thermoset plastics are like an egg. Once boiled, an egg can no longer go back to its liquid form. Thermoset plastics are 'cured' during manufacture via a chemical reaction that can occur at room temperature or when heat and pres-sure are applied. Thermoset plastics cannot melt; they simply burn. Examples of thermoset plastics used in jewelry production are: Bakelite-types (phenolics and ureas), caseins, and epoxies. [39]

Thermoplastic
Thermoplastics are like an ice cube. Water can be used in its liquid state or it can be frozen as ice; it can then be liquefied again and reused. Thermoplastics are heated to a molten state to form. Once cooled, they hold the new shape, but can be re-melted and re-formed. Thermoplastics are often shaped by the process of injection molding. Examples of thermoplastics used in jewelry pro-duction are: Celluloid-types (cellulose nitrate), acrylics (Lucite & Plexiglas), cel-lulose acetates, and polystyrenes.[40]

Thermoforming
Thermoforming is the process where a thermoplastic piece is heated-up for a second shaping to add greater detail or for blow-molding. Blow-molding is a process where steam or heated air is forced between sheets of plastic mate-rial, which are set into a mold. The material softens and conforms to a mold. Blow-molding adds a three-dimensional quality and is more commonly seen in the manufacture of toys and packaging.[41]

Nut charm bracelet in green polystyrene. Unsigned. c.1950s. $35-40.

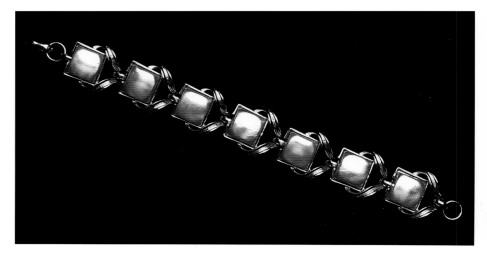

Bracelet of lavender, coated polystyrene squares. The coating created an artificial luster on the otherwise dull-colored polystyrene. Unsigned (A Coro knock-off). c.1950s. $25-30.

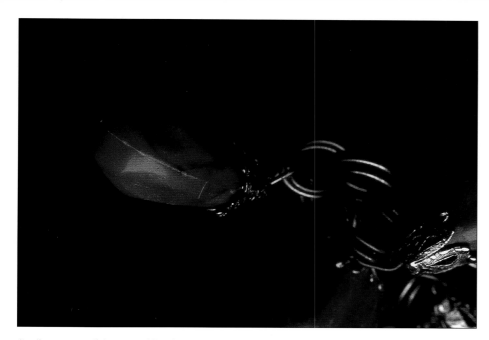

An illustration of the typical hard-parting line on jewelry of polystyrene composition.

Methods of Identifying Plastics

By examining the color, texture, smell, and even the sound of your jewelry, you can learn clues about the plastic's composition. The following tests can be performed to aid in this process. This is a comprehensive list of tests; some are invasive and costly.

The Rub & Smell Test

Smell tests are an easy, non-destructive way of determining a plastic's composition. With your thumb and forefinger, rub a piece of plastic jewelry briskly for five to ten seconds, then smell. The smells emitted by plastics will vary in strength due to the purity of the chemicals used in the manufacture of the piece and the combination of chemicals used in the manufacturing process, as well as pigments or fillers added to provide color and texture.[42]

The smells various plastics emit:

Celluloid-type (cellulose nitrate or pyroxylin) —Camphor
Bakelite-type (phenolic) — Formaldehyde
Thioureas —Sulphur
Ureas — Fishy
Galalith-type (casein) — Burned Milk
Cellulose Acetate — Vinegar
Lucite-type (acrylic) — No Smell
Polystyrene — No Smell
Epoxy — No Smell

The Hot Water Test

Hot water is another method for bringing out tell-tale smells in plastics; however, only jewelry which is entirely plastic (i.e. no metal parts or rhinestones) should be run under hot tap water and not for more than 30 seconds. This method is most appropriate for testing solid plastic bangle bracelets. Rhinestones, painted-on detail, and old glued-in settings will be ruined if they become wet. Cellulose acetate should not be immersed in hot water, as it will distort.[43, 44]

Float It

Acrylic plastics will float in water, while the Bakelite-type will sink. Thermosets are usually denser than thermoplastics, and almost all thermosets are more dense than pure water. This test takes advantage of the diffences in the specific gravity of the various plastics.[45] This method is most appropriate for plastic jewelry that contains no metal parts, rhinestones, or painted-on detail. It is suggested that one use hot water in a clean quart jar, then add sugar and salt in such a combination that it registers 1.25 on a battery tester. This recipe is the best combination to achieve accurate testing. The salt and sugar make the water denser; therefore, most thermoplastics will float on it. [46]

Chemical Testing with Household Cleaners & Polishes

Certain household polishes and cleaners can indicate whether your jewelry has a formaldehyde base. This would include plastics of the phenolic (Bakelite, Marblette), urea and thiourea (Beetle), or casein (Galalith) varieties. The formaldehyde present in these plastics will react with the following commercially available substances: Formula 409, Metal-Glo, and Wenol metal polishes. Formula 409 and Metal-Glo will leave a yellow residue on a cloth, while Wenol will leave an orange residue if the item being tested has formaldehyde its composition. Never use a metal polish on your jewelry that has a gritty formulation, as it could scratch your jewelry. Use a small amount of product on a soft cloth or cotton swab, and always test in an inconspicuous spot, as finish can be rubbed-off.[47, 48]

Listen

Gently, very gently, tap two pieces of Bakelite-type plastic together to hear a dull sound. Tapping two acrylic pieces together produces a higher-pitched sound. The acoustic properties of plastics are related to its stiffness and density.[49, 50]

Touch

Glass has a cold feel while plastics feel warm. A simple, unscientific test to determine whether your jewelry is composed of glass or plastic parts is to hold the component of the piece up to your lips or face.

Invasive & Expensive Testing Methods

Pinning

Pinning is an unacceptable form of jewelry testing because it involves destroying your jewelry. Pinning will only show whether your jewelry is thermoset or thermoplastic. A heated pin will easily slip into thermoplastic materials, while the pin will not penetrate a thermoset piece. This method is dangerous for testing Celluloid-type plastics (cellulose nitrate), as it is highly flammable and may react violently upon contact with a red-hot pin.[51]

Testing with Solvents

Certain solvents can dissolve plastics. Like heat, solvents can also have a damaging effect upon thermoplastics. Obviously, this is not an ideal testing method because it is so destructive. Thermoset plastics (phenolics, ureas, thioureas, caseins, and epoxies) don't dissolve. Acetone will dissolve Celluloid (cellulose nitrate), cellulose acetates, and acrylics, while benzene will dissolve polystyrene. On the back of the plastic piece, place a small drop of the solvent; you will see evidence of the substance dissolving the plastic.[52, 53]

Laboratory Testing

Analytical testing using a Fourier Transform Infrared (FTIR) Spectrometer will absolutely tell you the type of plastic used in your jewelry; however, it cannot tell you the specific brand name of the plastics used (i.e. Catalin, Marblette, Plexiglas). Unfortunately, it can cost you several hundred dollars per item tested. The machine itself can be yours for a mere $30,000. The machine shines a beam of infrared light on the plastic and measures which wavelengths are absorbed or reflected. The resulting spectral "fingerprint" is then compared to a library of chemical fingerprints and, voila, you will know with certainty whether you are dealing with phenolic, acrylic, polystyrene, etc. The machine's one drawback is that when testing a painted or coated piece, it will key-in on the spectral fingerprint of the paint or coating. Painted or coated jewelry from the mid-century, however, often has a polystyrene base anyway. A FTIR Spectrometer can be found at a private contract laboratory or materials science department of a university.[54]

Some Basic Visual Clues

Acrylic plastics are color stable. Old pieces will retain their true colors, while Bakelite-type and epoxy pieces will change color and yellow with age. Urea formaldehydes have brighter colors than their molded and cast phenolic cousins.[55] Polystyrenes, often used in the manufacture of inexpensive jewelry, have a hard parting line. Although the parting line from the mold can be tumbled and buffed-off, polystyrenes can be further characterized by their hard, dull, brittle nature. Sometimes polystyrenes were coated to give them the artificial luster that they lack. Acrylic pieces were traditionally more costly to manufacture than pieces produced in cellulose acetate or polystyrene; therefore, look at the overall quality of the piece when trying to determine its chemical make-up.[56] Telling the difference between the acrylic and the cellulose acetate-type plastics is the most difficult, as they share similar properties of color and texture. Cellulose acetate can be successfully produced in a thinner form than the acrylics. Acrylics become brittle when they are too thin; therefore, they are usually seen in thicker plastic pieces. Celluloid-type plastics, too, can be produced in very thin sheets.[57] When determining the composition of your plastic piece, consider the age of the object. Although there are overlaps in the timeline of plastics manufacture, you can often tell the approximate date by the style of the piece and, thereby, the basic composition of the material used. Occasionally, more than one type of plastic is used in the manufacture of costume jewelry. Often, Bakelite-type pieces were combined with acrylics. So, in these cases, the jewelry uses plastic parts of both the themoset and thermoplastic types. Sometimes, the plastics used were blends of two different materials (i.e., a Lucite® and styrene mixture) (See Coro Moonray ad, page 25). In the case of a hybrid plastic, basic testing would be difficult. Often, a combination of testing methods may be needed to determine the type of plastic used. Remember that the jewelry you are testing is most likely between 40 and 80 years old, so rough testing methods will destroy your treasures. Enjoy the beauty of the piece and simply appreciate the chemistry that went into its manufacture.

How Parts Were Supplied & Molded

Thermoset plastic pieces made from molding powders are often formed using heat and pressure, while liquid thermoset resins are typically cast into a mold to cure. Time and temperature are critical in making a thermoset product.[58] Bakelite-type jewelry was often carved from rods and tubes of cured, cast phenolic. Bracelets and rings were sliced off the rod and then jig sawed and carved to make the finished product.[59, 60] Molded pieces are generally very dark in color and can contain filler substances such as wood flour to make the objects more pliable.[61] Molded phenolic for industrial uses may have contained other fillers like cotton for increased impact resistance or asbestos for increased heat resistance. The Bakelite Company had literally thousands of formulas for their molding compounds, which were customized for specific applications or customers.

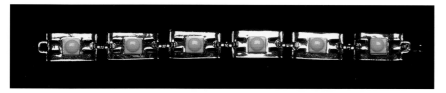

Bracelet back illustrating the molding mark of injection molded plastic. Usually, the molding mark is obscured by the jewelry hardware.

Magnificent example of a figural pin utilizing both Bakelite and Lucite-type materials. Unsigned. c.1940s. Jewelry courtesy Studio V, Chicago. $175-200.

Thermoplastics are generally injection molded from pieces that look like pellets or granules. Celluloid-type plastic could not be successfully injection molded, due to its flammability.[62] The plastic pellets are melted under high heat and pressure and then forced into a cooled mold. The material quickly hardens and the mold is opened to eject the object. Fillers can be also added to thermoplastics. A common example of this would be mica in thermoplastic pieces to provide a pearly look.[63]

The first injection molding machines were produced in Germany by H. Bucholz in 1919. In 1922, the Grotelite Company of Madison, Indiana was the first American company to import these injection molding machines. Grote Industries is still in business and produces automobile headlamps. Eckert and Ziegler, another German Company, began producing injection molding machines in 1927[64] and, in 1930, the Foster Grant Company purchased one of these machines. By the mid-1930s, Foster Grant was the among the first American companies to produce injection molded plastic jewelry.[65]

The Suppliers of Molded Plastic Parts

Companies specialized in making plastic parts used in costume jewelry manufacturing. Coro had 16 of their own plastic molding machines, but also used outside plastic molders and suppliers. Brier Manufacturing (Little Nemo) had sixteen of their own molding machines as well. Most other jewelry companies did not have the volume of in-house molding machines so they outsourced or imported their plastic products. Several of the leading American companies that manufactured plastic jewelry parts were Ace Plastics, Best Plastics, and Melmar Plastics. Sobel Brothers specialized in plastic confetti or fleck jewelry parts. All of these manufacturers were located in the New York and New Jersey areas and, sadly, all now are out of business.[66, 67, 68]

Leominster, Massachusetts, the plastics capital of the United States, had many companies that manufactured plastic jewelry and jewelry parts, including The Foster Grant Company.[69, 70] Norm Kaufman, former President of First in Imports, was an importer of plastic jewelry parts. He sold many tons of plastic pieces to Coro. Coro was the largest company and could demand that anything be sold to them on an exclusive basis, Kaufman said, "The importers went to Coro first…then they went to the others." He did add, however, that the same shaped plastic part could be sold to others, but in a different color than what Coro had chosen.[71] This is the reason, perhaps, that we find jewelry with the same shaped plastic parts both signed and unsigned, as they reasonably could have come from different companies.

Besides rhinestones, many of the parts for plastic jewelry came from Austria, Czechoslovakia, and Germany. Some of the more notable companies from the Kaufbeuren or Neu Gablonz region of Germany include the Bartel and Emil Hubner companies. A company by the name of Walter Fischer specialized in plastic leaves. This area of Germany was often called the "bead district" and was formed after World War II. The original area of production for rhinestones and glass beads was Gablonz, in the Bohemia area of Czechoslovakia, but it was decimated by the war and the communist take-over of Eastern Europe. After World War II, much of the rhinestone manufacture moved to Austria. Many of the large manufacturers, including Coro and Brier Manufacturing (Little Nemo), had offices in this region prior to World War II that closed with the onset of the war in Europe.

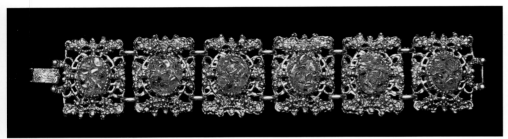

This green bracelet illustrates the plastic granules used in thermoplastic molding. In this instance, the plastic pellets used in this bracelet are simply fused together and kept intact. c. 1960s. Unsigned. $35-40.

Other regions notable for the manufacture of plastic jewelry and jewelry parts in Europe are Oyonnax, France and Milan, Italy (and its surrounding area). In Asia, Osaka, Japan was known as a center for costume jewelry, particularly plastics. The plastics industry was among the first industries in Japan to return to operation immediately after World War II, helping to buoy their economy. [72, 73, 74]

Swarovski & the AB Rhinestone

Although this book is about plastic jewelry, a major event in the history of the rhinestone must be mentioned because it is helpful in dating jewelry. In 1956, Manfred Swarovski, of the crystal company Swarovski in Austria, developed "Aurora Borealis" (AB) glass rhinestones, in cooperation with the design house of Christian Dior. AB rhinestones are coated with thin metal particles before being steamed in a vacuum to create a distinctive iridescent luster. These extra sparkly stones are often used in the production of costume jewelry. [75]

Manufacturing Process

Designs

Costume jewelry must start out with a design. The economic depression of the 1930s caused many unemployed fine jewelry designers to find work in costume jewelry companies. [76] For example, in 1930 Alfred Philippe joined Trifari

after having done fine jewelry design for Cartier and Van Cleef & Arpels. [77] Designs were crafted by in-house design teams, talented company owners, jobber factory employees, free-lance designers, and even Rhode Island School of Design students. [78, 79, 80] Although there have been many notable and famous costume jewelry designers (Design Director Adolph Katz and Chief Designer Gene Verrecchia (Gene Verri) of Coro, Kenneth Jay Lane, and Lea Stein, to name a few), much fabulous costume jewelry is unsigned. We may never know the names of the talented individuals who created many masterpieces.

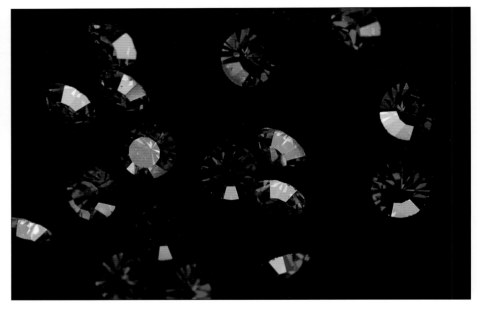

Dazzling, iridescent Aurora Borealis (AB) rhinestones.

The design of costume jewelry is generally influenced by current fashion trends. "Costume jewelry is based on the fashion of the times," according to Jack Feibelman, former Director of Product Development for Coro. He added that the designers often copied the Parisian fashions of the day, as their styles were one season ahead of the United States. Feibelman further notes that a successful item produced by Coro would have had a production run of nearly 150,000 pieces. Costume jewelry was often produced in sets, most commonly comprised of a necklace, bracelet, earrings, and a pin. Sometimes a ring or second style of a bracelet or earrings might also comprise the complete set. A complete set is called a parure; a partial set is called a demi-parure. Often, a set was manufactured in several colors. Feibelman said that Coro usually produced sets in six different colors. [81] Although some costume jewelry companies, like

Trifari, copied fine jewelry, according to Jim Axelrad, former owner of Pakula, "The jewelry design was based on the color of the season to accent and enhance the wardrobe." "A season is sixteen weeks," Axelrad continues. Costume jewelry, Axelrad further notes, has a shelf-life of twenty-six weeks or less. Axelrad now works with computers and compares the similarities in the two industries. "I have my best design minds making what you have now obsolete," Axelrad said.[82, 83] We know, however, that much of the magnificent design of vintage costume jewelry is actually timeless.

Modeling and Molding

From the paper design for a piece of jewelry, a model is created. In the case of a manipulated or strung and beaded piece, jewelry parts are pieced together like a puzzle. However, in the case of a piece that requires casting, a model is actually sculpted from a block of metal.[84, 85]

Historically, the Trifari Company's sales materials best describe the modeling process. "Finally the model is completed. It is tediously and carefully and entirely made by hand. It looks exactly like the un-plated finished piece of jewelry will look. To the expert eye, however, it is a trifle thicker. This allows for natural shrinkage later on when the model is used to make a mold."[86]

The model details the exact design, down to the rhinestones or plastic parts to be used. The model can be used to fine-tune the design, can determine what parts will be needed for the production of the finished product, or can determine structural weaknesses in the piece of jewelry before production begins. After the design is sketched and the model created, a mold must be crafted in order to create multiples of the individual piece.[87]

The molding process is also excellently detailed in old Trifari Company sales manuals. "[The model]…is carefully laid out on a bed of sand so fine it feels like powder. With tools as delicate as those used by a dentist, a mold room man embeds the model into the sand until each contour around edge, each hole in the piece, is completely packed with sand. The model must not be pushed too far down—it must not lie too high—it must be just so. For where the sand level stops, that is where the two edges of the mold will eventually meet. Now on top of that half of the model not covered by sand, a liquid plaster is poured. We now have the model buried in a mold, half of which is sand and half of which is plaster. In about twenty minutes the plaster hardens and the sand is washed away. Over this plaster half is placed a thick mat of rubber, and slowly under heat and pressure, the rubber mat pressing against the plaster mold and the exposed piece of the model become half of a rubber mold. The plaster half of the mold is broken away leaving the model in the half rubber mold. In another hour and a half of heat and pressure in a vulcanizing machine, the second half of the mold is made."[88]

HOW JEWELRY IS MADE ....

Design:
An artistic drawing is made by one of our talented designers.

Blueprint:
A blueprint of the piece is made indicating size, proportion and dimension.

Model:
A model is sculptured to look like the actual piece.

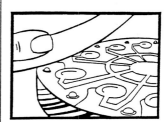

Mold:
A mold is made from the model.

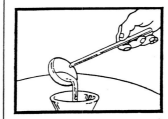

Casting:
The piece is then cast, which means the material being used is poured into the mold and allowed to harden. The material could be pewter, brass or plastic, depending upon the piece.

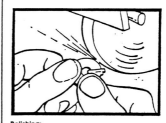

Polishing:
Each piece is polished several times to smooth out the roughness and eliminate scratches.

Electroplating:
Metal pieces are electroplated with copper, nickel and then either gold, silver or rhodium to achieve the desired creative look.

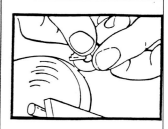

Finishing:
Based on the desired creative look, a final high gloss polish or satin buffing is selected.

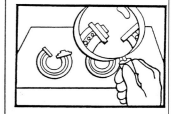

Stones:
Gem quality stones are added by hand and the piece is assembled into its final form.

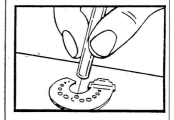

Epoxy:
This is the step that accommodates adding fashion color to each piece. Epoxy paint is applied by hand using an airgun.

Packaging:
Care is taken to safely package our jewelry. Each piece of jewelry is individually wrapped before it is shipped to your store.

T R I F A R I ™

Trifari design and manufacturing diagram from Trifari company sales materials. c. 1970s. Courtesy Wendi Mancini.

In a large operation like Trifari, more than one mold would be produced, as continual use would wear out the mold. Molds could also be made of bronze.[89, 90]

Casting and Plating

The next operation in producing a piece of costume jewelry is casting the piece, usually in a base metal like brass (copper & zinc alloy) or white metal (lead & tin alloy). A casting machine is a drum-like container, where the halves of the mold are inserted and clamped into place. When the machine is closed, the mold spins and molten metal is then poured into a spout in the lid. Through centrifugal force, the metal enters each cavity of the mold. The machine continues to spin the mold until the metal hardens. Once finished, the piece is removed from the mold and is placed in a vibratory tumbler which does the final cleaning, preliminary polishing, and prepares the piece for electroplating. Mold marks (gates), stray pieces of metal, or rough spots are removed in this tumbling process. The findings (i.e. earring backs, clasps, and pin-backs) are then soldered on to the castings.[91, 92]

Women arranging earrings on racks at the in preparation for plating. Antonelli Plating Company, Providence, Rhode Island, c.1950s. Photo courtesy of The Providence Jewelry Museum.

The jewelry at this stage still lacks its luster and shine, as it needs to be plated. Costume jewelry is often plated with gold or rhodium. Rhodium, a member of the platinum family, is generally preferable to silver, as it does not tarnish. Electroplating is a very exacting process that is accomplished by wiring the piece of jewelry to a conductive rack with a strand of fine copper. This wiring of the metal jewelry parts gives it a positive electrical connection to a source of current and allows the plating metal to adhere to the base metal. First, undercoating metals (like copper and nickel) are plated onto the base metals so that the final coating will be smooth and will adhere properly. The jewelry goes through a series of chemical washes and baths in order to prepare for the final electroplate of the chosen metal. Each step in the process is carefully timed and the chemicals must be kept pure and at their proper concentrations. The timing and temperature of the chemical washes and baths, plus the amount of electricity and agitation used, can produce very different results.[93, 94, 95] Trifari had an independent plating division that provided electroplating for outside companies.[96] The hazardous waste water generated by electroplating processes contains chemical contaminants, heavy metals, and acids; therefore, strict EPA & OSHA regulations regarding the handling and disposal of effluent have driven much of the plating operations off-shore.[97]

Finishing

In the final production steps, pieces are polished and buffed to a high shine, rhinestones and plastic parts are either glued-in or prong-set, and the jewelry is carded, paper-tagged, or boxed for shipment to stores.[98]

Providence: A Guild Town

The people who work at jewelry companies are highly skilled. Jane Civins, of the Providence Jewelry Museum, has stated that at the beginning of the last century, jewelry manufacturing personnel actually met the incoming emigrant boats at Ellis Island, New York, to search for workers with jewelry-making expertise. They would take qualified new immigrants and their families back to Providence, Rhode Island, and give them work. The predominant ethnic groups working in Providence costume jewelry industries were Jewish, Italian, and Portuguese.[99] Many of the factories were family-owned operations, and within the factories many workers came from the same family. Generations of some families specialized in a single operation or skill, such as enamelling or producing metal filigree.

In the early 20th century, there were hundreds of "job shops" in Providence. Small, usually family-owned and operated, these factories might provide a single function for other manufacturers, such as casting, linking, or soldering. In the same way that specialty goods were manufactured in Europe, Providence became a "guild" town.[100, 101] Although there were labor unions in Providence, they did not have the strength of those in the New York area. Some jewelry manufacturers moved their operations to Providence because of the guild atmosphere.[102]

Jewelry making was hard and labor-intensive work. The same effort went into a costume piece as that of fine jewelry.[103] "Homework," a common practice among jewelry workers, were goods that were finished at home when workers were paid on a per-piece basis. Women and children primarily did the "homework" that usually involved the jewelry's assembly or the carding (mounting the jewelry on sales cards). Today, the practice of "homework" is illegal in Rhode Island.[104]

Women assemble plastic watches and flower pins, Providence, Rhode Island, 1950. Company unknown. Photo courtesy of The Providence Jewelry Museum.

Wavy leaf, turquoise parure by Leru with original hang-tag. c. 1950s. Jewelry courtesy of Cheryl Killmer. $75-85.

Now, due to prohibitive labor costs in the United States, much of the jewelry production has moved to China, Taiwan, and Thailand. Surprisingly, about the time that many of the major costume jewelry manufacturers were closing and consolidating, the industry hit its peak of employment. In 1978, 32,500 of Providence's manufacturing jobs were in the jewelry industry. As of 2004, only 7,900 jewelry workers remain.[105]

Jim Axelrad, the former owner of jewelry manufacturer Pakula, has stated that the industry's peak in the late 1970s was due to an influx of new customers, as women of the baby boom generation reached their twenties. He also belives this was the time jewelry manufacturing began to move overseas. The popular beaded and puka shell jewelry of this era was imported to the United States.[106] Karl Eisenberg, of Eisenberg Ice, also attributes the peak of employment for Rhode Island jewelry workers to jewelry trends and fashion magazine promotions in the late 1970s of fad items such as mood rings and stick-pins.[107]

Who Made What for Whom

Because of the many specialized jewelry factories, it can be difficult to tell who actually made what. Although some of the largest companies did much of the manufacturing in-house (Coro, Trifari, and Brier [Little Nemo]), others did not need factories because excellent jewelry manufacturing resources were available in Providence. Lisner and Pakula, for example, did not have their own factories for much of their existence and Kramer never had its own manufacturing facility. They relied on the job shops to do their manufacturing. Many of the job shops never made jewelry under their own banner, but produced jewelry solely for others. Iraida Garey, former Vice-President of Product Development for Lisner, has said that if she and Lisner owner Victor Ganz decided they wanted a nautical look, they could go to a factory in Providence that specialized in this and, with the factory owner, develop a nautical line that would bear the Lisner name. Sometimes several jobbers worked on the same piece. Jim Axelrad has stated that seven or eight job shop contractors worked on Pakula's pieces. You could find a jobber for any aspect of a specific operation, from casting to gluing, soldering, polishing, plating, carding, and wrapping.[108, 109, 110, 111]

Patents, Copyrights & Lawsuits

Although it was common for many jobbers to be involved in the manufacture of an individual piece from beginning to end, the ownership of the design generally belonged to the company that commissioned the jewelry. Copying jewelry designs was rampant, but the larger companies had the resources to protect their designs through copyrights and patents.[112] In 1955, Trifari won a landmark federal U. S. copyright case, Trifari, Krussman & Fishel, Inc. vs. Charel Co., Inc., allowing them to copyright fashion jewelry design as a work of art. The court's opinion in this lawsuit states, "Characterizing plaintiff's product disparagingly as 'junk jewelry' defendant urges that it not rise to the dignity of a 'work of art.'" The opinion further states "…a piece of costume jewelry is entitled to copyright protection and generally that right is not dependent upon judicial appraisal of its artistic merit. A necklace, like a circus poster or book, is not to be denied the benefits of the Copyright Act because it may not attain the same recognition as is accorded the work of a renowned artist."[113] Sometimes companies put a copyright or trademark symbol on a piece, but never actually went through with the paperwork, as it was cost-prohibitive to formally register each design. Jim Axelrad of Pakula says that occasionally when he brought his product to the plater and polisher, he would receive back several pieces less than the gross he brought in. The explanation from the plater and polisher was that the pieces ended up at the bottom of the tank. However, later, when Axelrad attended jewelry shows, he saw that his designs had been copied.[114]

Trifari patent of a link chain for a necklace, 1954. Courtesy of Jim Katz, http://www.KatzJewelers.com.

Growing Demand for Costume Jewelry in the Depression Years

Not only the fine jewelry designers warmed to fashion jewelry market during the 1930s, the buying public, looked to costume jewelry to refresh their tired clothes when they were unable to afford fine jewelry. A 10-cent pin from Woolworth's could make an old dress look new.[115, 116]

The 1940s

The costume jewelry business in the United States boomed during World War II, despite shortages of raw materials for jewelry making. When the supply of rhinestones from Czechoslovakia and Austria was curtailed by the war,[117] the Ralph Singer Company (Ora) made some jewelry with plastic parts.[118] Even by 1946, shortages forced manufacturers to meet only fifty-percent of their orders.[119] According to Karl Eisenberg, the owner of Eisenberg Ice, "Our business during the war was unbelievable because of the fact that at that time women were working and were making good money." The makers of costume jewelry had to use sterling silver in war time when pot metal, which is largely made of tin, was requisitioned by the U.S. military. The U.S. government also required that the jewelry materials be accurately marked. "By law it was required to have 'Sterling' on it," Eisenberg said.[120]

Providence's Major Contribution

During World War II, many of the jewelry factories in Providence converted their manufacturing plants to making specialized and precision parts for the war effort. Some companies produced military insignias while others made military hardware.[121] Trifari, for example, made bullet parts and dies for Remington Arms, torpedo parts for the U. S. Navy, and even Royal Air Force emblems. Trifari donated the proceeds of these emblems to England's Royal Air Force. Trifari is most famous for their ingenious use of the Lucite scrap from the canopies of bombers for magnificent "Jelly Belly" jewelry![122, 123]

Distribution Process

The Sales Rep

Major jewelry companies depended on an army of sales representatives to distribute their jewelry in the early 20th century. Men, primarily, sold their product lines in specific territories of the country to large and small department stores and boutiques across the United States. In 1958, Wendi Mancini, who began work at Trifari as a secretary and showroom sales associate, became their first-ever saleswoman representatve. Her male counterparts jokingly referred to her as "Bruce.".[124] "This is not due to any physical shortcoming, but rather because of my being the only female in a national sales force of 27, 'I am just one of the boys at Trifari'," Mancini said. She has stated her opinion that the lack of women sales reps was not due to flagrant chauvinism, but the physical realities of the job. "Lifting sample cases requires a strong back, and not many ladies are conditioned to rolling 200 lbs. of jewelry from car to hotel to store to home base," Mancini said.[125] Jim Axelrad, the former owner of Pakula, has said that his 32 to 35 sales reps had to carry twelve telescoping sample cases of merchandise and that many of them bought motor homes to accommodate this load.[126] Sales people could be on the road for 6 to 8 weeks at a time.[127]

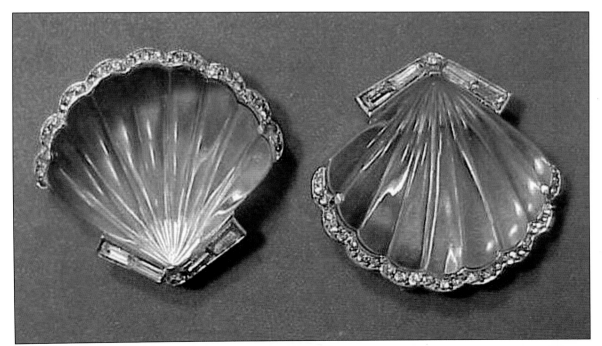

Jelly Belly fur clips by Trifari. Trifari called these unusual 1949 fur clips "Moonshells". Jewelry and photo courtesy of Pat Seal, Treasures from Yesterday, Fort Worth, Texas. $275-350 Pair.

Led by famed Executive Vice President Royal Marcher, Coro's dynamic sales force, of about 55 reps, was legendary.[128] According to Axelrad, the Coro reps had presence, "they were all image, all confidence."[129] Lisner had a small sales force of about ten reps throughout the country.[130] Many of the jewelry companies' salespeople were 'independent reps,' meaning they worked solely on a commission basis and had to front their own expenses. Mancini not only paid the salary for her assistant, but the travel expenses to and from the trade shows. A small number of costume jewelry companies sold their wares with reps that did home party plans, like Sarah Coventry.[131]

Department Store Buyers

The people on the other side of the sales transactions were department store buyers. Their job was to purchase merchandise for the upcoming season on behalf of the department store or boutique. Reps visited these buyers on-site, and the buyers attended four trade shows per year in New York City. The most important of these trade shows were in January and June.[132] According to Alan Marcher, a former salesman for both Coro and Lisner (and a nephew of Royal Marcher), "The mark-up on jewelry was fifty percent back then."[133] To-day, costume jewelry is sold to department stores on a guaranteed sale basis. If the merchandise doesn't sell, the department store sends it back to the manufacturer. According to Jim Axelrad of Pakula, two former Coro salesmen who went out on their own, Arthur Mayer of K&M Jewelry and Bob Koch of RN Koch, devised the 'guaranteed sale'. "They took all the risk away [for the department store]. If two to five percent doesn't sell, the department store gets credit for the old inventory," Axelrad said.[134]

Showrooms and Sales Offices

Most major costume jewelry companies had non-manufacturing executive offices and showrooms in New York City, plus showrooms, sales offices, and stock storage facilities in major U.S. cities like Chicago, Dallas, and Los Angeles.[135] Pakula concentrated on the small, hard-to-reach markets that many of the larger companies ignored. "We sold on the main streets of America," Axelrad said. He had a salesman who did a million dollars worth of business with small department stores in the Dakotas.[136] This proves that costume jewelry has a universal appeal; women from New York to Fargo *must* have their costume jewelry!

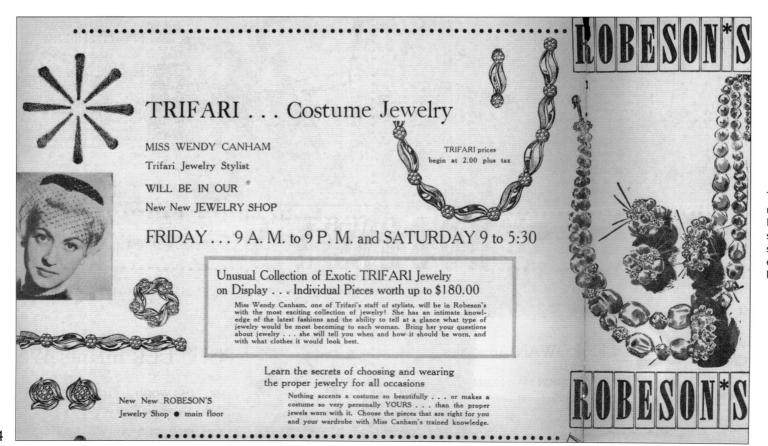

Trifari's first woman sales representative, Wendi Canham Mancini. Ad for Trifari trunk show at Robeson's department store (Champaign, Illinois), c. early 1960s. Courtesy of Wendi Mancini.

2.
Selected Manufacturers

Although there were many fine manufacturers of plastic costume jewelry, we will concentrate on three of the most important and prolific companies and two other smaller manufacturers that produced unique objects of great beauty.

Coro

Coro hang-tag featuring Pegasus.

A Coro hallmark.

Coro Ad "Spring-into-Summer", 1956. Note the "Moonray" squares, a signature Coro product from the 1950s. According to Jack Feibelman, former Director of Product Development for Coro, "Moonray" actually refers to the type of plastic used in the jewelry's manufacture, not the style. "Moonray" was a Lucite® and styrene mixture, while other common monikers in Coro ads such as "Moonbeam" were comprised of 100% Lucite® and "Moonglow" of powdered thermoset resin. Plastic blends, like the Lucite and styrene mix, often yield an opalescent appearance. Ad courtesy of Pat Seal, Treasures from Yesterday, Fort Worth, Texas.

"Coro is King"

Coro was, by far, the largest American manufacturer of costume and plastic jewelry. "Coro is King" was the motto used by department store buyers and even reps of the rival companies.[137] In 1902, Emanuel (Manny) Cohn, an importer and salesman of costume jewelry with an existing company called E. Cohn & Company, and sales agent Carl Rosenberger merged to form the new company of Cohn and Rosenberger.[138, 139] Although Cohn died in 1910, his name remained part of their moniker.[140] In 1943, the corporate name Coro was officially adopted from the first two letters of each of the founders' last names. Originally manufacturing in New York City, Cohn & Rosenberger established their first factory in Providence, Rhode Island, in 1910.[141]

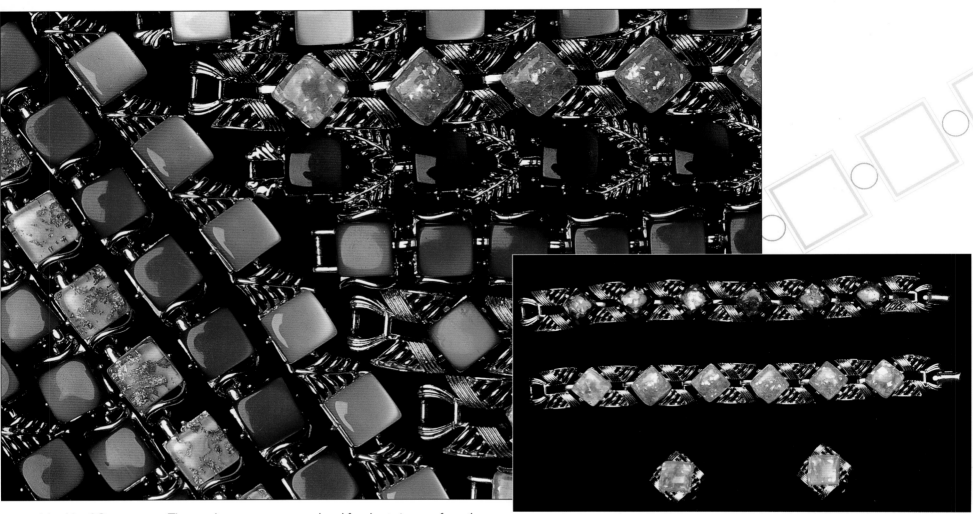

A jumble of Coro squares. The popular squares were produced for about six years from the mid-1950s into the early1960s in about a dozen new colors a year to accent spring and fall fashions. Pastels for spring and darker colors for fall. That makes over 70 variations!

Burgundy and white confetti examples of the famous Coro squares. c. 1950s. Bracelets $35-40, each. Earrings $15-20.

Royal purple parure of Coro squares. c. 1950s.
Necklace, bracelet, and earrings $60-75 set.

Emerald green demi-parure of Coro squares. c. 1950s. Necklace
courtesy of Judith Levin. Necklace and bracelet $50-65 set.

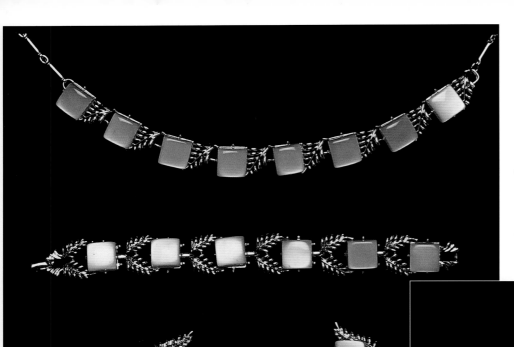

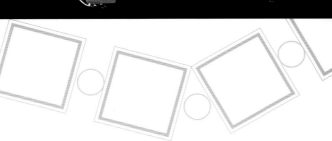

Pale pink parure (Say that three times real fast!) of Coro squares. c. 1950s. Necklace, bracelet, and earrings $60-75 set.

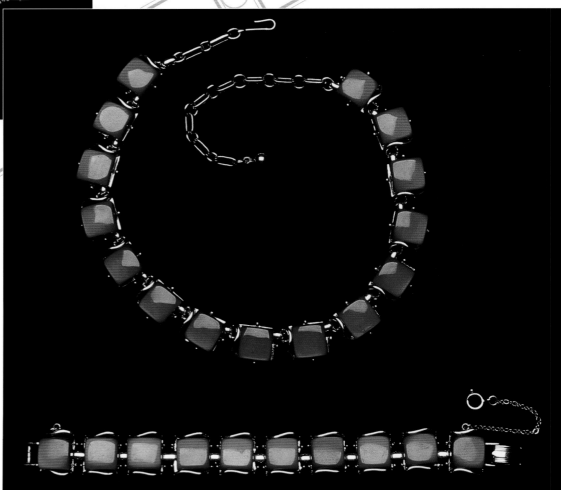

Cherry red Coro squares, demi-parure. c. 1950s. Necklace and bracelet $50-65 set.

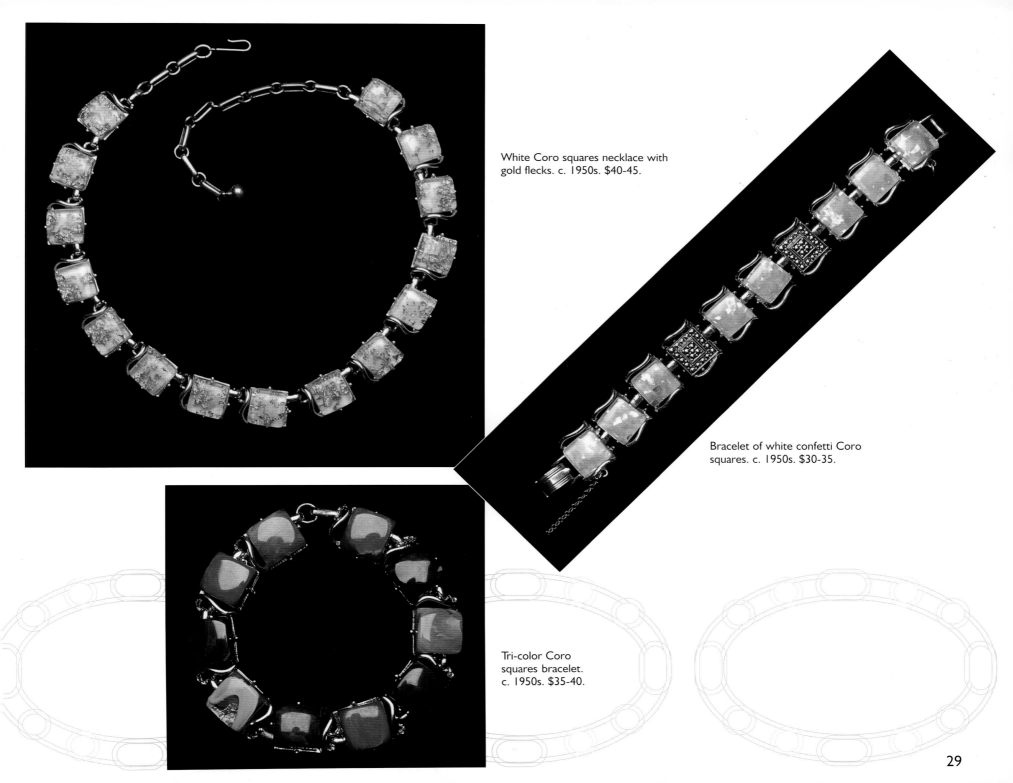

White Coro squares necklace with gold flecks. c. 1950s. $40-45.

Bracelet of white confetti Coro squares. c. 1950s. $30-35.

Tri-color Coro squares bracelet. c. 1950s. $35-40.

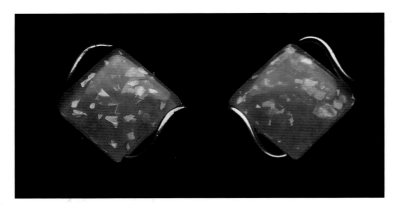

Large size squares (3/4" square), white Coro confetti earrings. Coro also made a small variation (1/2" square) of this earring. c. 1950s. Jewelry courtesy of Judith Levin. $15-20, each.

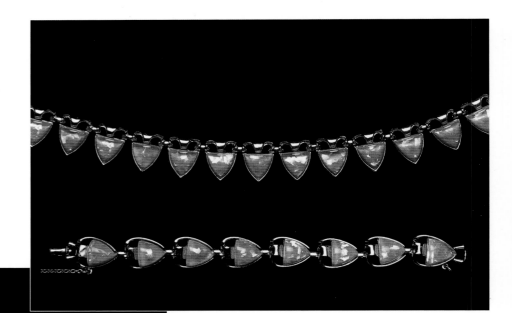

Triangular-shaped, Coro confetti necklace and bracelet in pink. c. 1950s. Set $55-65.

Coro confetti dots. c. 1950s. Necklace in white $25-30 and bracelet in blue $30-35.

Pink Coro bracelet with half-circle shapes. c. 1950s. $30-35.

A twin of the pink bracelet. Gold fleck version of this half-circle, shaped Coro bracelet. c. 1950s. $35-40.

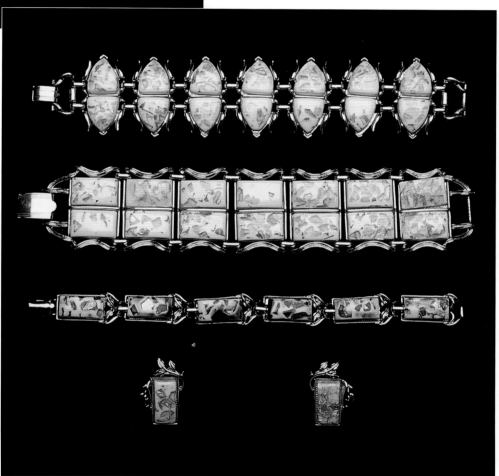

A variety of Coro composition fleck pieces. c.1950s. Cream bracelet with gold flecks $40-50. Cream bracelet with pink and gold flecks $40-50. Cream bracelet with black and gold flecks $30-35. Cream earrings with pink flecks $15-20.

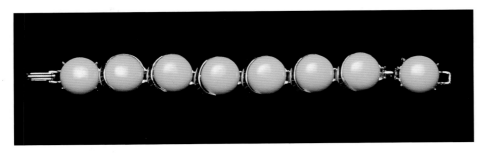

Coro bracelet composed of bright yellow dot shapes. c. 1950s. $30-35.

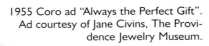

1955 Coro ad "Always the Perfect Gift".
Ad courtesy of Jane Civins, The Providence Jewelry Museum.

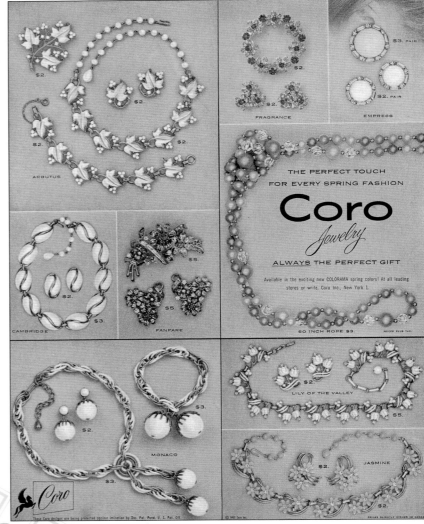

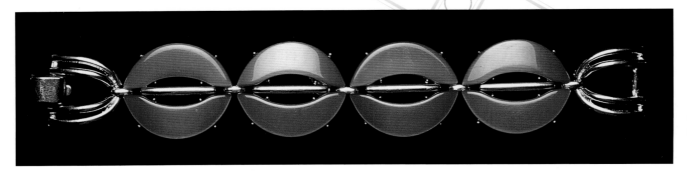

Mod Coro bracelet in purple and raspberry. c.1960s. $45-55.

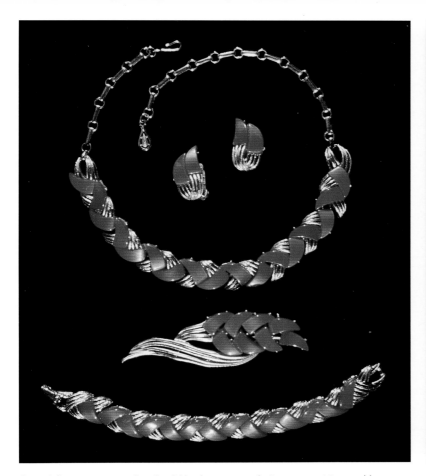

Coro full parure in a red and gold basket-weave design, comprising necklace, bracelet, pin, and earrings. c.1950s. Jewelry and photo courtesy of The Francesca Medrano Collection. Set $125-135.

Coro blue, confetti parure. Set comprising necklace, bracelet, and earrings on original card. A never worn piece of costume jewelry with their intact, original cards or hang-tags can increase value. c. 1950s. Jewelry and photo courtesy of The Francesca Medrano Collection. Set $85-95.

Coro "Cambridge" bracelet and earrings in olive green. 1955. Bracelet $35-40. Earrings $15-20.

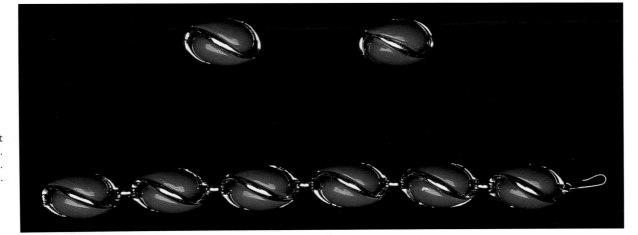

Coro offered public issues of common stock, first in 1929 and later in 1945. The 1929 stock issue was used to build the largest manufacturing plant of that time in the costume jewelry business. Although the plant initially stood empty, it positioned Coro for the great demand for costume jewelry that developed in the latter part of the 1930s and for the subsequent wartime production needs in the mid-1940s. The 1945 stock offering was used to enlarge the plant further, for which the company's foresight benefited with the boom of post-war consumerism.[142, 143]

The Coro company became a worldwide concern, with branches and factories in Toronto, Canada (1920s); London, England (1927), and Taxco, Mexico (1942).[144] Stateside, sales and stock storage offices were established in New York, Chicago, Los Angeles, San Francisco, Miami, Dallas, and Atlanta.[145] Coro produced jewelry that sold to both higher-end (the Corocraft & Vendome lines) and lower priced jewelry.[146] By 1946, Coro produced 16 percent of all costume jewelry made in the United States.

In 1957, Carl Rosenberger died at the age of 85. His son, Gerald Rosenberger, succeeded him as president, but died suddenly of a heart attack in January 1967. The decline of Coro had already begun when Executive Vice President Royal Marcher retired in 1960.[147] In 1969, the conglomerate Richton International Corp., headed by Franc M. Ricciardi, bought 51 percent of Coro's stock, and bought the remaining shares in 1970.[148, 149] (Richton once owned the Oscar de la Renta brand.[150]) By 1979, Richton closed Coro's U.S. facilities and filed for Chapter 11 bankruptcy protection in March of 1980.[151, 152, 153] Only Coro Canada, Inc., based in Toronto, and its U.K. holding company, Richton International, Ltd., remained after the company emerged from bankruptcy in 1981. In 1988, Richton sold their U.K. fashion jewelry subsidiary to the Swarovski group. Richton operated their Canadian facility until 1992, when it was sold to a Venezuelan investment group that included the distributor of Coro's fashion jewelry products in Latin America.[154] The Venezuelan group seems to have run the company until the late 1990s. Their last known address was in Montreal, Quebec, and according to information provided by the Bureau du Commercial, Quebec, the last Dunn and Bradstreet economic report available on Coro Canada Inc. dates from 1997.[155] Richton is still in business and concentrates on the distribution of lawn sprinklers and fountains.[156]

Trifari

"The Three Kings"

The three kings of Trifari were Gustavo (Gus) Trifari, Sr., Leo F. Krussman, and Carl M. Fishel. In New York City in 1918, Gustavo Trifari, Sr. was designing and manufacturing elaborate, rhinestone-encrusted hair ornaments when he met salesman Leo F. Krussman, who worked for a hair ornament house. Times were changing; women were cutting their hair and making elaborate hair ornaments obsolete. The two men saw instead a future in costume jewelry and formed a partnership. Trifari designed and manufactured pins, while Krussman sold them. The business was a success and in 1925, Carl M. Fishel, a former president of a large hair ornament business, joined Trifari and Krussman in the venture.[157, 158] The three men's strengths built a very successful company: Mr. Trifari ran the factory, Mr. Krussman was in charge of finance, and Mr. Fishel led the sales force. Although commonly known as Trifari, its actual name was Trifari, Krussman and Fishel, Inc. According to former Trifari saleswoman Wendi Mancini, the Trifari family name was originally pronounced 'Treh-fairy'. When Hallmark Cards purchased the company in the late 1970s, they changed the pronunciation to 'Treh-fahry', as they felt it was easier to say.[159]

A Trifari Hallmark.

Trifari Ad "Punctuation White" 1960.

Classic Trifari necklace of blue "V's". c. 1950s. Jewelry courtesy of Cheryl Killmer. $25-35.

Three elegant, white Trifari bracelets. One with a matching pair of earrings. c.1950s. Bracelet with egg-shaped pieces $20-25. Bracelet with bell-like parts $35-40. Bracelet and earrings with round, flower-shapes $40-50 set.

In 1930, fine jewelry designer Alfred Philippe joined Trifari.[160] In 1946, Trifari jewelry generally sold for $10 and up, making it a higher-end costume jewelry purchase.[161] In 1952 and 1956, Alfred Phillipe designed costume jewelry for First Lady Mamie Eisenhower to wear at the inaugural balls.[162]

Advertising was an essential part of the Trifari program. In 1938, Trifari commenced national advertising and department store clients received two percent of the previous year's billing for co-op advertising. According to Irving Wolf, former president of Trifari, "It helped to build Trifari to what it was. Stores were impressed with the caliber of the advertising". According to Wolf, advertising throughout the year enabled the business to thrive all twelve months of the year, as opposed to typically doing three-quarters of the year's business in the last three months of the year.[163]

In 1939, due to problems with labor unions, Trifari moved all its manufacturing from New York City to Providence, Rhode Island. Management, design, and sales offices remained in New York City.[164, 165] Gustavo Trifari, Sr., died in 1952. Krussman and Fishel retired in 1964 and the sons of the original owners, Gustavo Trifari Jr., Louis F. Krussman, and Carlton M. Fishel, assumed their father's positions and responsibilities. In 1975, Trifari was sold to Hallmark Cards, and in 1988 Hallmark Cards sold Trifari to Crystal Brands.[166] Crystal Brands renamed themselves the Monet Group and, in 2000, the Monet Group, including the Trifari brand, became part of Liz Claiborne, Inc.[167]

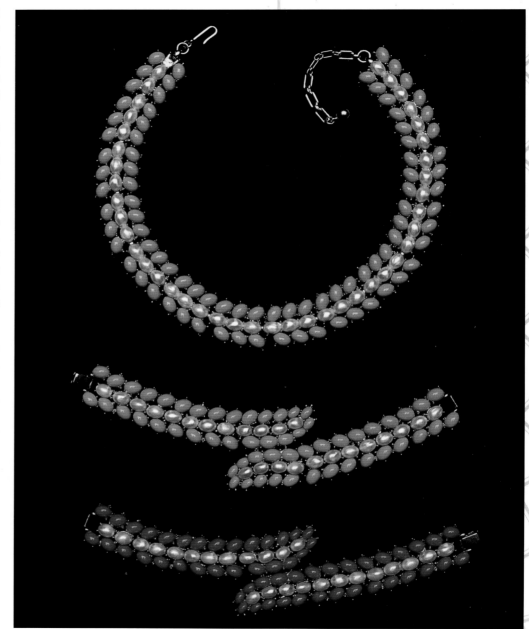

Trifari combines pearly, plastic cabochons with others in coral and light blue for these lovely Native-American-inspired jewels. c. 1950s. Blue set courtesy of Cheryl Killmer. Necklace $45-55. Bracelets $45-55, each.

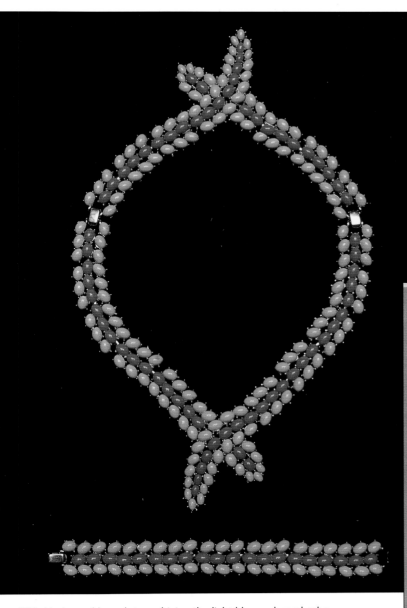

Trifari lariat and bracelet combining the light blue and coral cabochons. c. 1950s. Jewelry courtesy of Cheryl Killmer. $125-150 set.

Magnificent, white Trifari lariat. c.1950s. Jewelry courtesy of Cheryl Killmer. $55-65.

Trifari Ad "She will Think it's Christmas!" 1953.

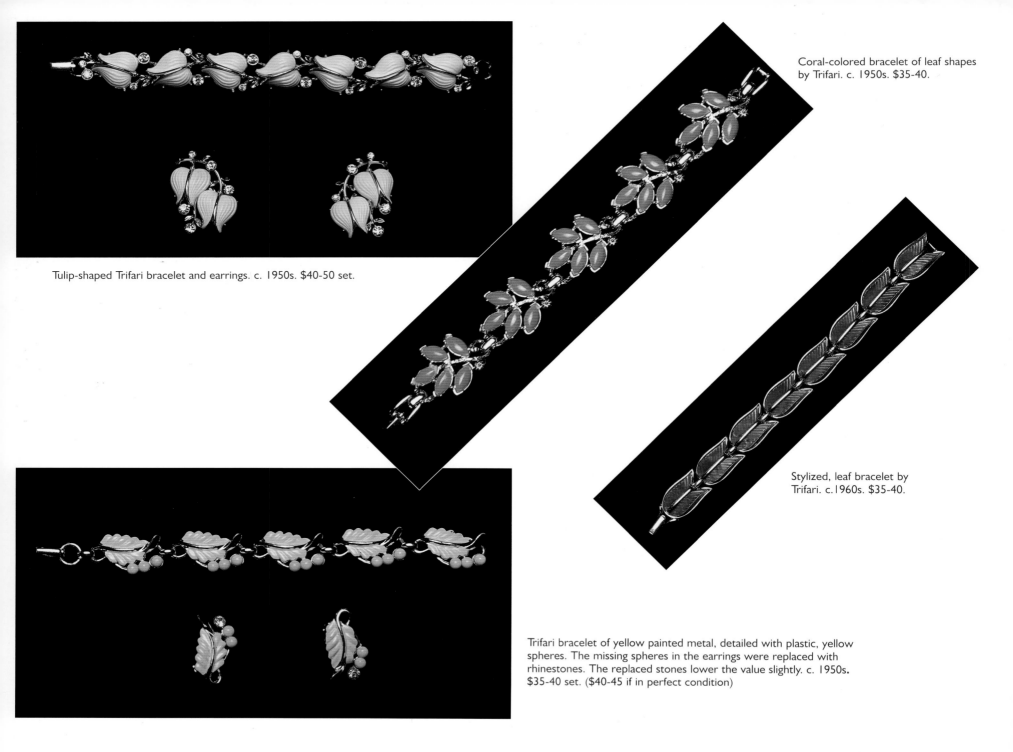

Coral-colored bracelet of leaf shapes by Trifari. c. 1950s. $35-40.

Tulip-shaped Trifari bracelet and earrings. c. 1950s. $40-50 set.

Stylized, leaf bracelet by Trifari. c.1960s. $35-40.

Trifari bracelet of yellow painted metal, detailed with plastic, yellow spheres. The missing spheres in the earrings were replaced with rhinestones. The replaced stones lower the value slightly. c. 1950s. $35-40 set. ($40-45 if in perfect condition)

The epitome of Trifari elegance. Light blue parure, which includes necklace, bracelet, and earrings. c. 1950s. Jewelry courtesy of Cheryl Killmer. Set $85-95.

Pink perfection. Trifari parure, comprising necklace bracelet, and earrings. c. 1950s. Jewelry courtesy of Cheryl Killmer. Set $85-95.

Lisner

D. Lisner and Co. was founded in 1904, by David Lisner. Lisner's son, Sidney Lisner, and a cousin, Saul Ganz, eventually joined the business.[168, 169] Until the 1930s, Lisner's business was largely wholesale and much of the jewelry was imported from Europe. Lisner was the U.S. importer and distributor for Schiaparelli jewelry. Lisner also promoted the Prince Machabelli perfume line and had other product lines that included clocks, crystal, giftware, and hatpins.

Urie Mandle (the father of Robert Mandle of R. Mandle Co.) joined Lisner as a full partner in the 1930s. His job was to build the jewelry line and produce jewelry domestically, in Providence, Rhode Island.[170] The Lisner identification mark was first used in 1938.[171]

Saul Ganz's son, Victor, joined the family business in 1934. Victor was a renaissance man, renowned for amassing the largest and finest collection of artworks by Pablo Picasso, including one of Picasso's best-known and greatest masterpieces, *The Dream*. His collection also included works by 20th century masters Jasper Johns, Eva Hesse, Roy Lichtenstein, Claes Oldenberg, Robert Rauschenberg, Frank Stella, and Cy Twombly.[172]

Magnificent Trifari parure with red and green cabochons. c. 1960s. Jewelry courtesy of Cheryl Killmer. Set $450-500.

Trifari ad from Marshall Field's department store flyer, 1955. Ad courtesy of Marshall Field's (Chicago, IL).

white laurel ensemble by Trifari

Composition white leaves encased in a rim of gold-plated base metal to wreathe your throat, wrist or ears . . . to complement a pretty summer costume. Collar-type necklace has comfortable, adjustable hook-closing, bracelet with strong safety catch is made to flex easily, matching earrings have spring-clip backs.

38 YT 72—adjustable collar-type necklace $8.25*
38 YT 71—flexible bracelet 5.50*
38 YT 70—white laurel earrings with clip backs ....pair, 3.30*

*including 10% federal excise tax
Jewelry—First Floor. Also Evanston, Oak Park, Lake Forest and Park Forest

17

Right:
Victor Ganz (1913-1987) poses with his one of his Picasso masterpieces "Seated Woman",1987. Photo courtesy of Kate Ganz Belin.

Lisner's exquisite design sensibility came from owner Victor Ganz's impeccable taste in art.

Lisner's main office was in New York City, at 393 5th Avenue. When one stepped off the elevator there, they could see an enormous Frank Stella painting hanging on Victor Ganz's office wall.[173] Ganz also served on the board of directors of New York's Whitney Museum of American Art.[174] After Ganz's death, at the age of 74, in 1987, a portion of his art collection was sold at a 1988 Sotheby's auction for forty-four million dollars. In November of 1997, after the death of Victor's wife, Sally, more of the art collection was sold at a Christie's auction and brought a record price for a single-owner sale of $206.5 million dollars.[175, 176]

Victor Ganz was involved in every aspect of Lisner jewelry production and he traveled weekly between New York and Providence to oversee the manufacturing process. His artistic and creative eye could be found in everything from the jewelry design, the retail packaging, and the even the advertising. Ganz developed unique packaging for his jewelry. One design involved a long tube where the jewelry looked like it was hanging in mid-air; the line was called "Suspense." One of his ads, featured in a *New York Times* Magazine section, involved a brown paper cone from which flowered jewelry emerged, like a tiny bouquet. "He tried to get away from the Mamie Eisenhower thing," said his daughter, Kate Ganz Belin.[177, 178]

Sidney Welicky was one of Lisner's in-house designers. After Welicky's retirement, Iraida Garey, Vice President of Product Development, and Ganz himself took over designing responsibilities. Garey recalls trips to Paris and Rome for design inspiration and trips to Kaufbeuren, Germany, for further product development of the season's collections. Garey said that a Lisner bracelet was even featured in a 1950s Del Monte fruit cocktail ad. The ad had a 'Carmen Miranda'-type theme with a bracelet showcasing various plastic, fruit shapes spilling out of the can.

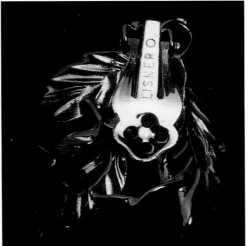

A Lisner hallmark.

In the 1950s and 1960s, the artistically designed Lisner jewelry retailed at an average of $3-5 per piece, making it a lower-end, but certainly not a lower quality, costume jewelry purchase.[179] In the mid-1970s, Lisner purchased the Richelieu Pearl Company from Joseph H. Meyers & Bros. and the company was re-titled the Lisner-Richelieu Corporation.[180, 181] Lisner-Richelieu was sold to Robert Andreoli of Victoria Creations in 1979.[182] In 1984, the parent company, Victoria Creations, was sold to Jonathan Logan. Jonathan Logan was then acquired by United Merchants and Manufacturers. In 1996, United Merchants and Manufacturers went into receivership and Andreoli purchased Victoria Creations back from the bank.[183] In 2000, Andreoli sold Victoria & Co., Ltd. to the Jones Apparel Group.[184] According to Andreoli, Lisner jewelry has not been manufactured since the mid-1980s.[185]

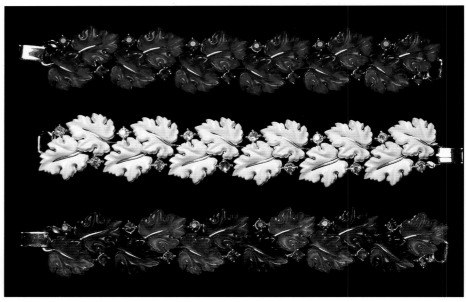

Lisner's most notable product: the resin oak leaves. The red oak leaf set is among the most coveted among collectors. The Lisner leaves were produced for about five years starting in the early 1960s. Bracelets: Red $75-100, white $35-40, blue $50-60.

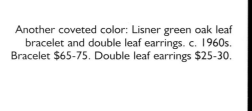

Lisner often used this unique decorative necklace extender charm.

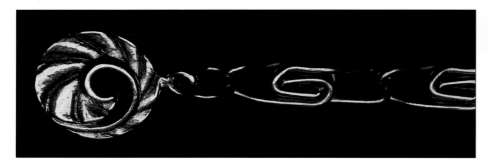

Another coveted color: Lisner green oak leaf bracelet and double leaf earrings. c. 1960s. Bracelet $65-75. Double leaf earrings $25-30.

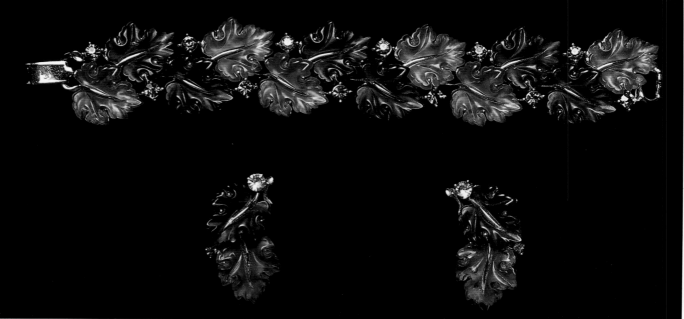

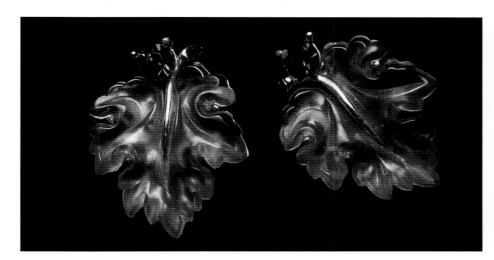

The single leaf version of the Lisner oak leaves earrings in brown. c. 1960s. Single leaf earrings $20-25.

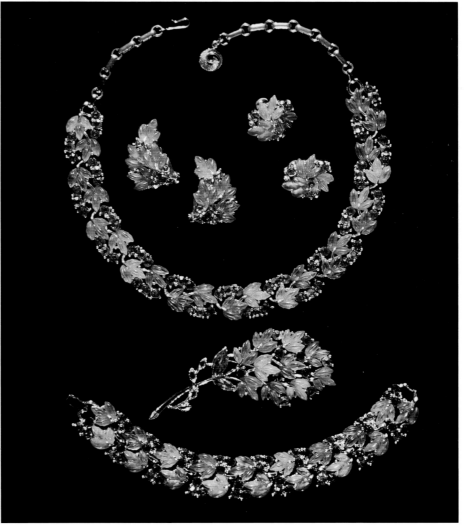

Magnificent Lisner full parure featuring small resin leaves. Notice the two variations of the earrings in this set. Entire parure c. 1960s. Jewelry and photo courtesy of The Francesca Medrano Collection. Francesca Medrano refers to this non-opaque type of plastic used in the Lisner leaves as "Jellies". The name truly fits! Set $175-195.

An unusual color in the Lisner oak leaves design. Pale yellow necklace and single leaf earrings accented with yellow rhinestones. There are numerous colors, color combinations, and variations in the Lisner leaf sets. Rare colors to watch for in the oak leaves style are smoky topaz and clear, completely clear, and lavender and purple combinations. c. 1960s. Jewelry courtesy of Cheryl Killmer. Necklace $50-60. Earrings $20-25.

Jelly Lisner heart-shaped, leaf bracelet and pin in brown, yellow, and orange. c. 1960s. Bracelet $30-35. Pin $35-40.

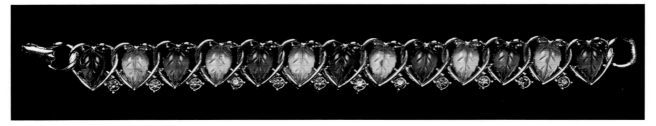

Green version of heart-shaped leaf bracelet. c.1960s. $30-35.

Lisner elongated leaf bracelet in cherry red. c. 1960s. $40-50.

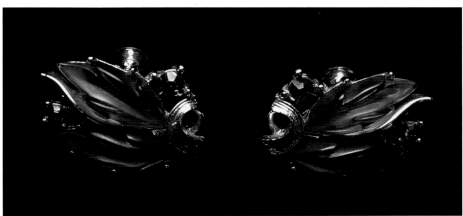

Green Lisner elongated leaf earrings. c. 1960s. $20-25.

Lisner large leaf earrings in yellow. Glue deterioration is often prominent with this leaf type. This set is in excellent condition. c.1960s. $30-35.

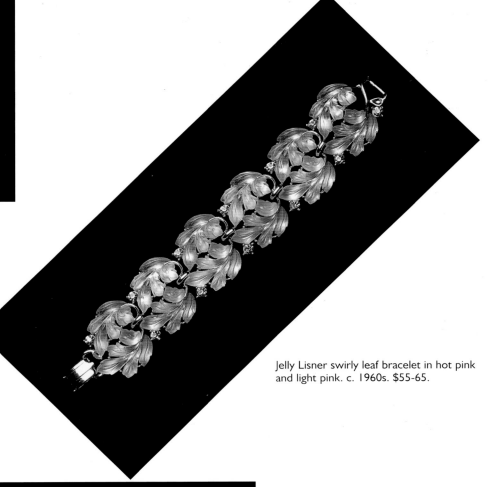

Jelly Lisner swirly leaf bracelet in hot pink and light pink. c. 1960s. $55-65.

Other color varieties of the Lisner large leaf. c. 1960s. Jewelry courtesy of Cheryl Killmer. Green pin $40-50. Green earrings $30-35. Tri-color, autumn-toned bracelet in $55-65.

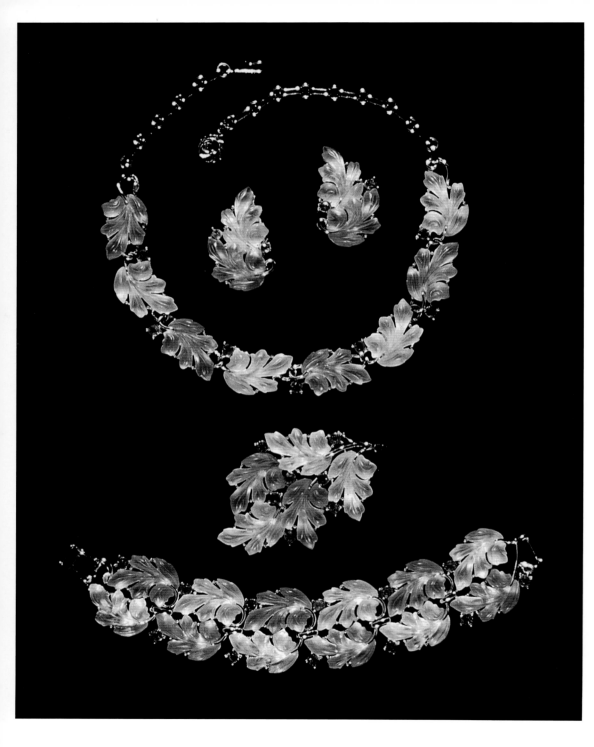

Lavender, Lisner swirly leaf full parure, composed of earrings, necklace, pin, and bracelet. This set also comes in a vibrant aqua color. c.1960s. Jewelry and photo courtesy of The Francesca Medrano Collection. Complete set $145-165.

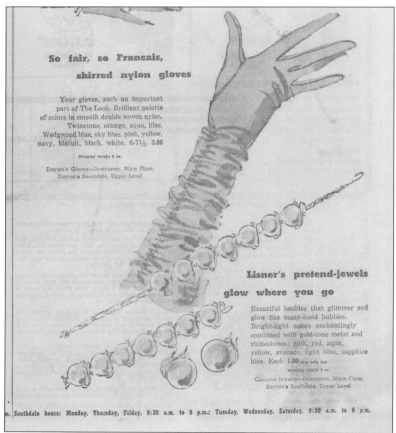

So fair, so Francais, shirred nylon gloves

Your gloves, such an important part of The Look. Brilliant palette of colors in smooth double woven nylon. Twinetone, orange, aqua, lilac, Wedgwood blue, sky blue, pink, yellow, navy, biscuit, black, white. 6-7½. 3.00

Shipping weight 8 oz.

Dayton's Gloves—Downtown, Main Floor; Dayton's Southdale, Upper Level

Lisner's pretend-jewels glow where you go

Beautiful baubles that glimmer and glow like many-hued bubbles. Bright-light colors enchantingly combined with gold-tone metal and rhinestones; pink, red, aqua, yellow, avocado, light blue, sapphire blue. Each 1.00 plus 10% tax

shipping weight 8 oz.

Costume Jewelry—Downtown, Main Floor; Dayton's Southdale, Upper Level

m. Southdale hours: Monday, Thursday, Friday, 9:30 a.m. to 9 p.m.; Tuesday, Wednesday, Saturday, 9:30 a.m. to 6 p.m.

"Lisner's pretend jewels glow where you go" from Dayton's department store (Minneapolis, MN) newspaper ad of January 14, 1957. Ad courtesy of Marshall Field's.

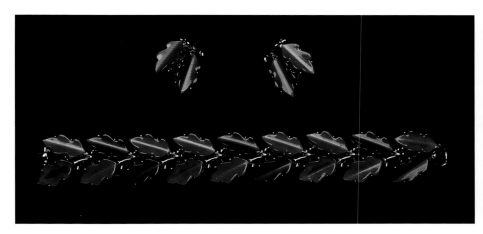

Lisner made lots of varieties of leaves. This set, in the tones of fall leaves, is opaque. c. 1960s. Jewelry courtesy of Cheryl Killmer. Earrings $15-20. Bracelet $40-45.

Perfect circles. Lisner bracelets in blue, red, and brown. ca.1960s. Each bracelet $35-40.

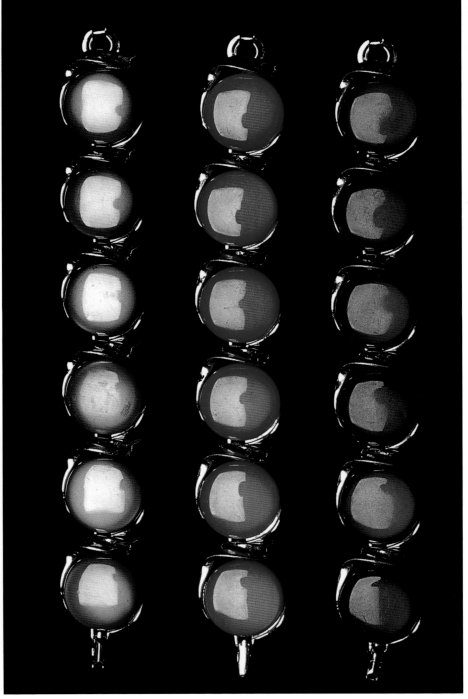

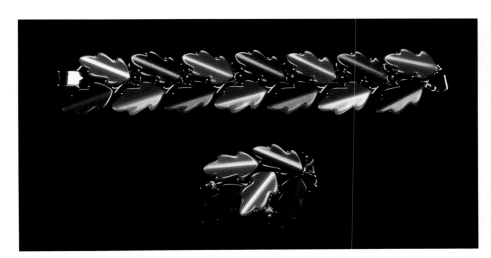

Blue version of the opaque Lisner leaf. c. 1960s. Bracelet $40-45. Pin has incredible detail with tiny, plastic acorns $40-45.

Artistic oblongs comprise these Lisner bracelets in blue and brown. c. 1950s. Each bracelet $40-50.

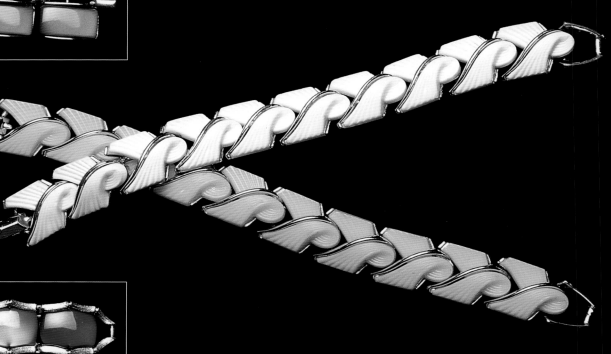

Deco-look Lisner bracelets in pink and yellow. c.1960s. Each $35-40.

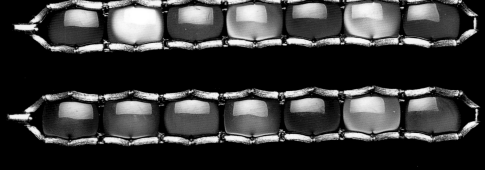

Asian-themed Lisner bracelets in blue and green show Victor Ganz's design sense. c.1960s. Each bracelet $40-50.

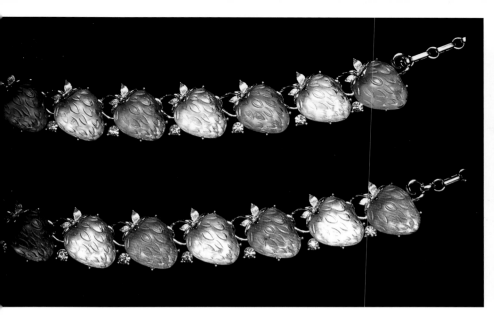

Highly sought-after Lisner jelly strawberries. Neon green and hot pink necklaces. The strawberries also can be found in blue. c. 1960s. Jewelry courtesy of Cheryl Killmer. Each necklace $75-100.

Aqua, Lisner opaque leaves. c. 1950s. Jewelry courtesy of Cheryl Killmer. Necklace $35-45. Earrings $15-20.

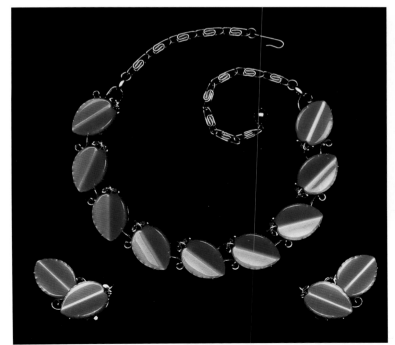

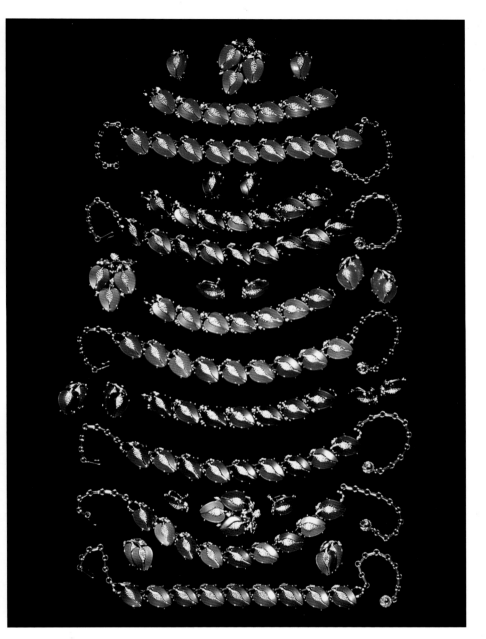

Lisner extravaganza. Stylized strawberry sets in red, green, orange, blue, brown, and purple. c. 1960s. Jewelry and photo courtesy of The Francesca Medrano Collection. Necklaces $35-40, each. Bracelets $35-40, each. Pins $30-35, each. Large earrings $20-25, each. Small earrings $15-20, each.

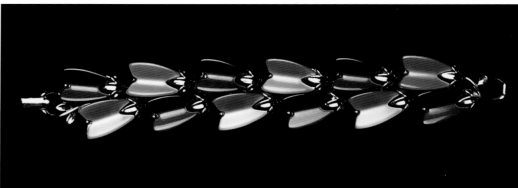

Abstract Lisner leaves in beige and brown bracelet. c.1950s. $35-40.

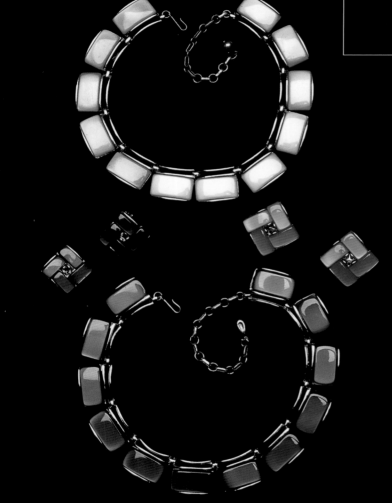

Sunburst Lisner parure. c.1950s. Jewelry and photo courtesy of The Francesca Medrano Collection. Set $135-155.

Blue-grey and aqua Lisner necklaces with matching earrings. Notice the size variety in the earrings. c. 1950s. Jewelry courtesy of Judy Levin. Necklaces $35-45, each. Earrings $15-20, each.

Kramer

Kramer Jewelry Creations was founded in New York City, in 1943, by three brothers: Louis, Harry, and Morris Kramer. Louis was the head of the firm.[186] It is surprising that this company, so prolific and that produced such high-quality and stylish merchandise, out-sourced all of their manufacturing.[187, 188] Kramer even produced pieces under a licensing agreement with Christian Dior, labeled "Dior by Kramer."[189] Lower-end Kramer pieces originally sold in the $5 range.[190] The company closed around 1980.[191]

A Kramer of New York hallmark.

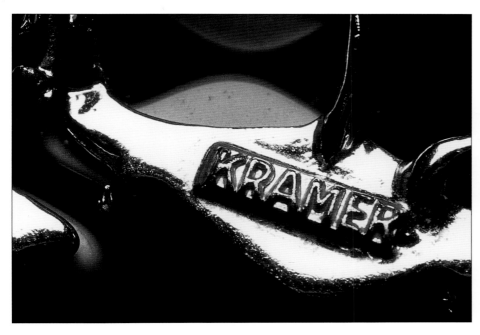

A Kramer hallmark.

Kramer white, star-shaped earrings and necklace of delicate, leaf-shapes. c. 1950s. Jewelry courtesy of Cheryl Killmer. Set $45-50.

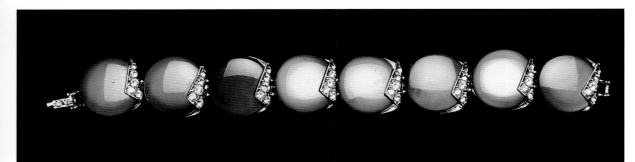

Brown Kramer bracelet
with rhinestone detail
and rhinestone-studded
clasp. c.1950s. $40-50.

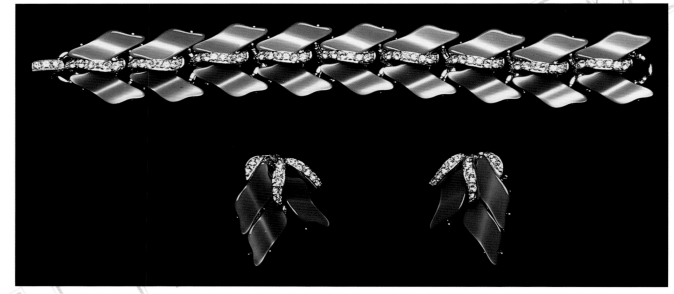

Elegant grey Kramer
bracelet and earrings with
rhinestone detail. c.1950s.
Earrings $20-25. Bracelet
$40-50.

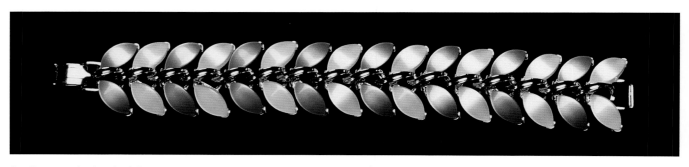

Small grey and white leaf shapes comprise this Kramer bracelet. c. 1950s. $35-40.

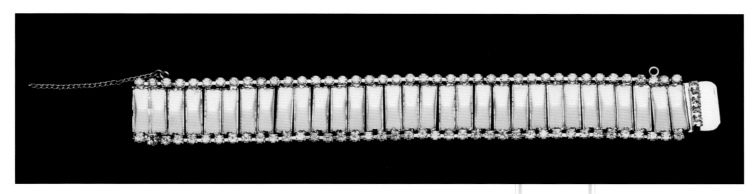

White Kramer of New York bracelet edged with clear rhinestones. c.1950s. $40-50.

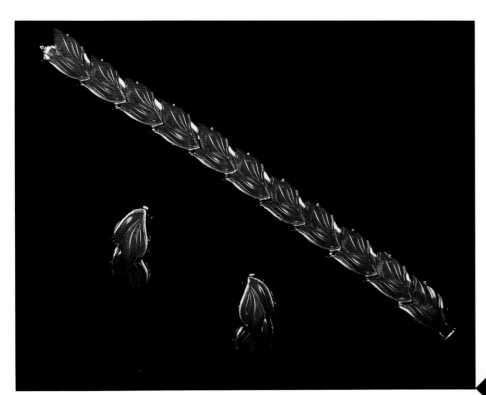

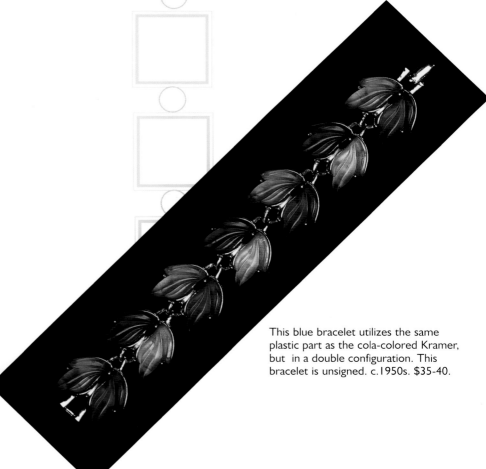

This blue bracelet utilizes the same plastic part as the cola-colored Kramer, but in a double configuration. This bracelet is unsigned. c.1950s. $35-40.

Cola-colored Kramer bracelet and earrings. Earrings are not marked. c. 1950s. Earrings $15-20. Bracelet $30-35.

Necklace and earrings in lavender and frost also utilize same plastic shape, but are also unsigned. Possibly Kramer. c. 1950s. Glue deterioration evident. This affects value. Earrings in current condition $5-10 (In excellent condition $15-20). Necklace in current condition $15-20 (In excellent condition $35-45).

Delicate Kramer parure in eggshell and beige, comprising necklace, earrings, and bracelet. c.1950s. Jewelry and photo courtesy of The Francesca Medrano Collection. Set $75-100.

Selro

Selro Corp. was founded by Paul Selinger, an independent jewelry representative for a number of manufacturers, such as Florenza and Capri. He served as a bridge between the factories and the wholesalers, and eventually sold under his own label. Selinger's fantastic jewelry was produced from the late 1940s into the 1960s. Sadly, little is known about this company that produced such exquisite costume jewelry.[192, 193]

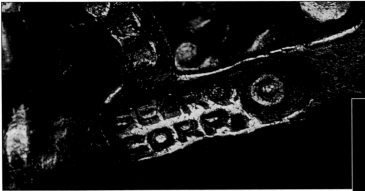

A Selro hallmark.

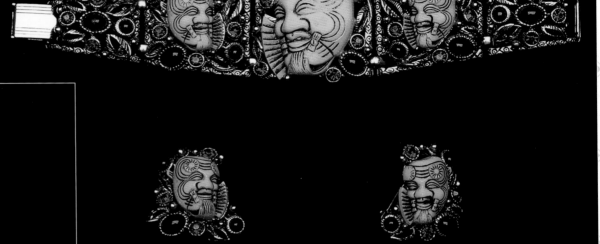

Brilliant yellow Selro (Paul Sellinger) Asian-faced figural bracelet and earrings. c. late1940s-early1960s. Jewelry courtesy of Cheryl Killmer. Bracelet and earrings set $250-295.

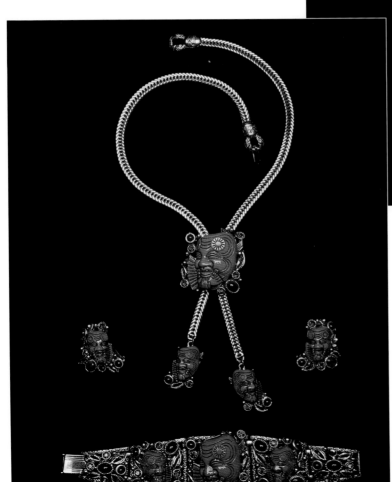

Bolo-style Selro necklace, bracelet, and earrings in deep coral. c. late1940s-early1960s. Jewelry courtesy of Cheryl Killmer. Necklace, bracelet, and earrings set $450-550.

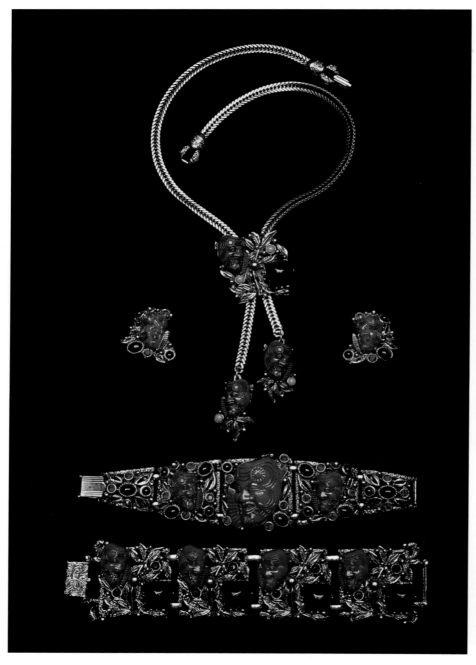

Red Selro faces comprise this set of bolo-style necklace, two bracelets, and earrings. c. late1940s-early1960s. Jewelry courtesy of Cheryl Killmer. Necklace $200-225. Each bracelet $175-200. Earrings $75-95.

White Selro captivating figurals, comprised of bolo-style necklace, two bracelets, and earrings. c. late1940s-early1960s. Jewelry courtesy of Cheryl Killmer. Necklace $200-225. Each bracelet $175-200. Earrings $75-95.

Bib-style, turquoise Selro necklace, with original hang-tag, plus bracelet, and earrings. The original price of the necklace was $22.50! c. late1940s-early1960s. Jewelry courtesy of Cheryl Killmer. Set of necklace, bracelet, and earrings $450-550.

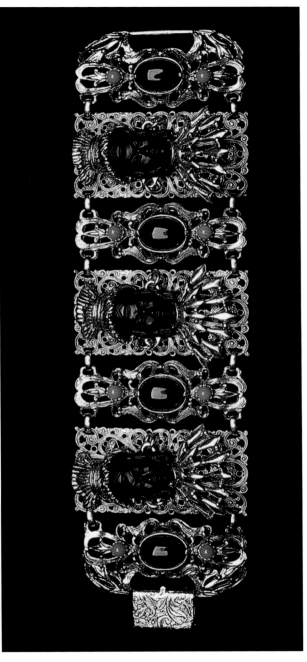

Bracelet with regal Selro blackamoors. c. late1940s-early1960s. Jewelry courtesy of Cheryl Killmer. $225-250.

More Amazing Figurals

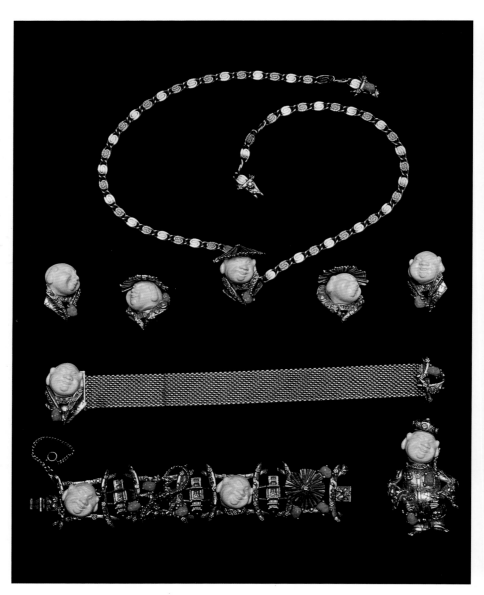

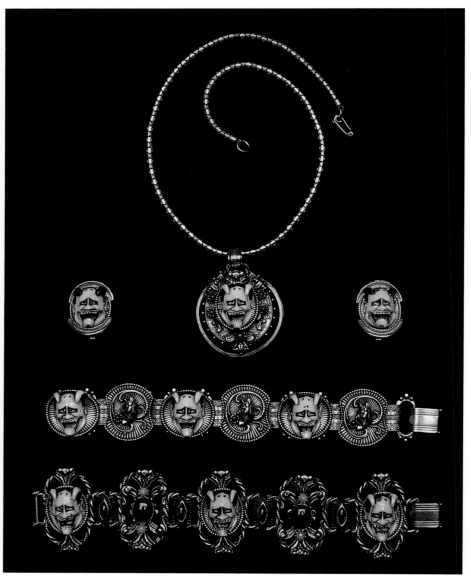

Another maker of fine figurals: Har (Harco Jewelry Co.) Chinaman set, including necklace, two bracelets, pin, and two sets of earrings. c. 1950s. Jewelry courtesy of Carol Sullivan. Complete set $900-1,200.

White devils set by Art (Arthur Pepper). c. 1950s. Jewelry courtesy of Cheryl Killmer. Necklace $75-95. Each bracelet $95-125. Earrings $30-35.

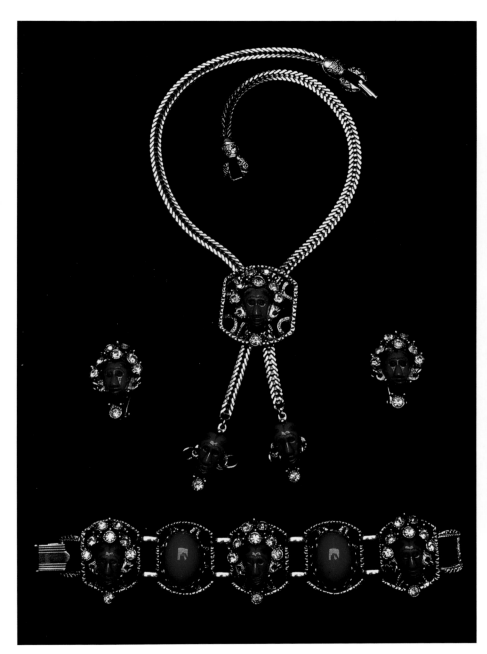

This unsigned figural set with creepy red faces still has great value. The necklace is very similar in to those in the Selro sets. c. 1950s. Jewelry courtesy of Cheryl Killmer. Set of necklace, bracelet, and earrings $175-195.

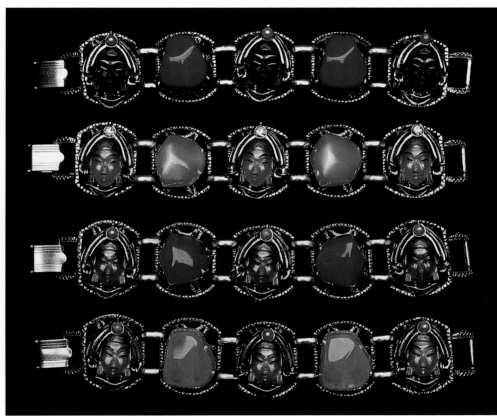

Four figural bracelets, unsigned in red, white, green, and yellow. Cheryl Killmer refers to this figural type as "Thai Girl". c. late 1940s to early 1950s. Jewelry courtesy of Cheryl Killmer. Each bracelet $75-95.

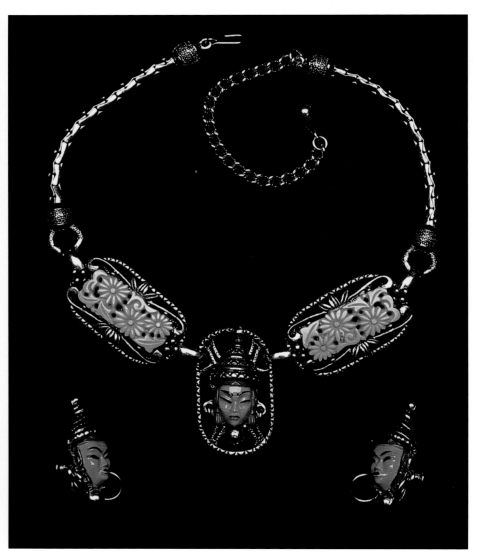

Turquoise 'Thai Girl' necklace and earrings, unsigned. c. late 1940s.
Jewelry courtesy of Cheryl Killmer. Necklace $75-95. Earrings $35-50.

Coral 'Thai Girl' necklace and earrings, unsigned. c. late 1940s.
Jewelry courtesy of Cheryl Killmer. Necklace $70-90. Earrings $35-50.

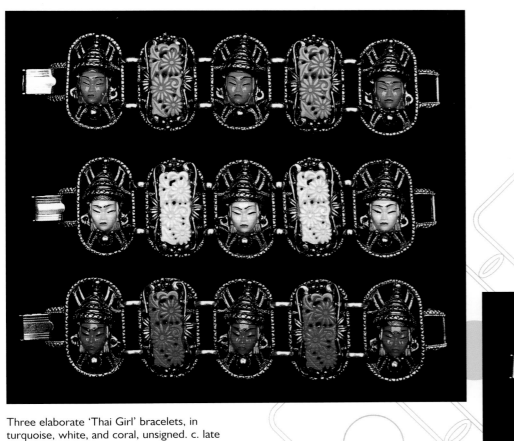

Three elaborate 'Thai Girl' bracelets, in turquoise, white, and coral, unsigned. c. late 1940s. Jewelry courtesy of Cheryl Killmer. Each bracelet $75-95.

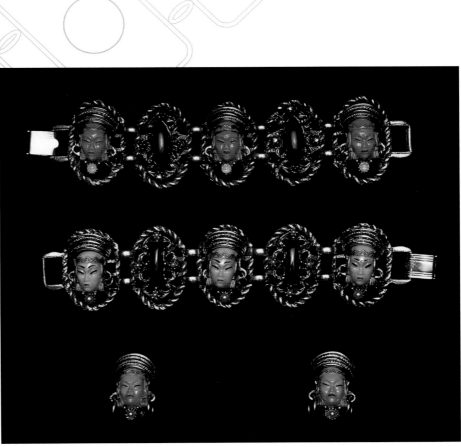

Two black and white 'Thai Girl' bracelets and set of earrings, unsigned. c. late 1940s. Jewelry courtesy of Cheryl Killmer. Each bracelet $70-90. Earrings $30-40.

Colors Galore

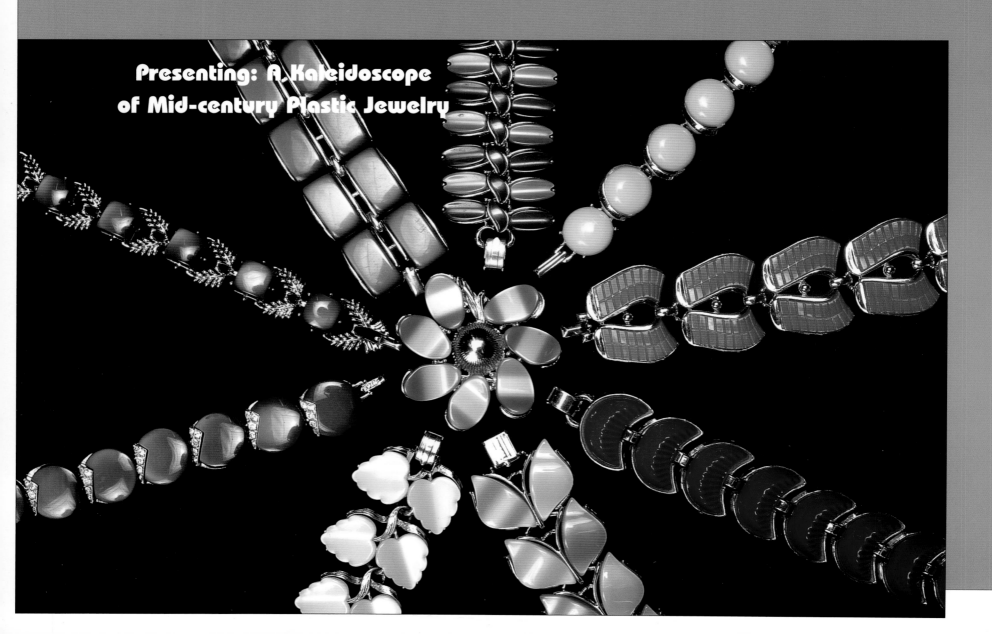

Presenting: A Kaleidoscope
of Mid-century Plastic Jewelry

Wonderful White

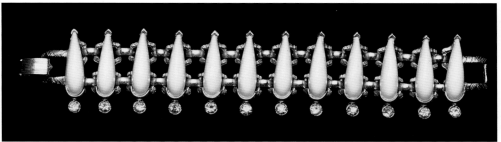

This bracelet looks like it has white exclamation points capped with clear rhinestones. Unsigned. $35-40.

Lightweight bracelet consists of white leaf-shapes, with a matching dangle. c. 1950s. Unsigned. $25-30.

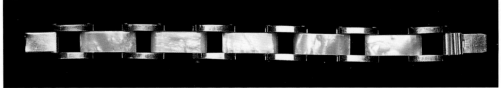

Deco bracelet with white, pearly , plastic oblongs. c. late 1940s. Unsigned. $20-25.

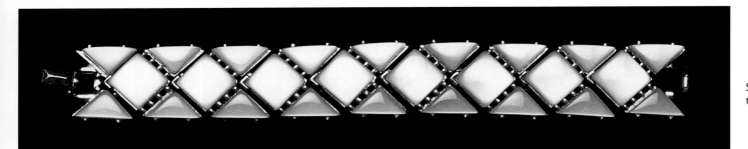

Silvery-white bracelet of squares and triangles. c. 1950s. Marked Charel. $35-40.

Monet bracelet with white wing-shaped parts. c.1960s. $35-40.

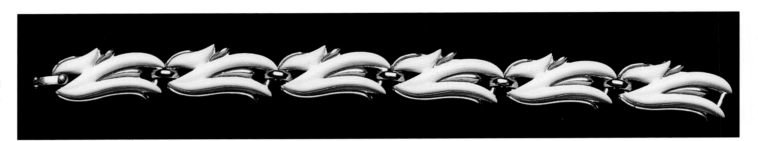

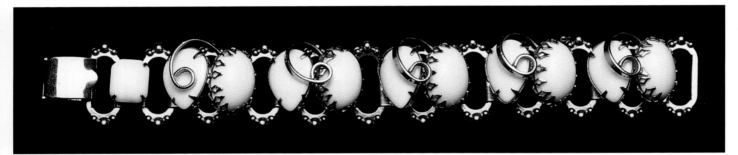

Bracelet of elaborate curly-cue hardware topped with white cabochon and teardrop plastic stones. Emerald-shaped stone at the clasp. c.1950s. Unsigned. $40-50.

Double-leaf and vine bracelet in white. c.1950s. Unsigned. $40-50.

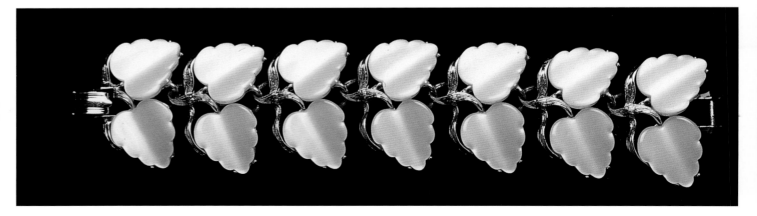

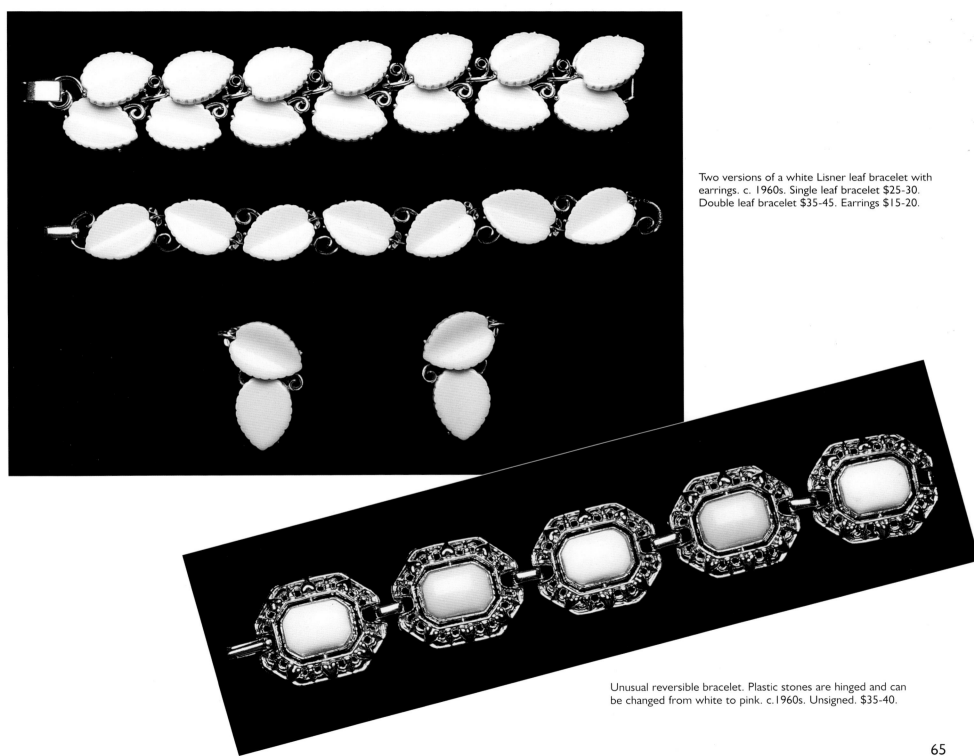

Two versions of a white Lisner leaf bracelet with earrings. c. 1960s. Single leaf bracelet $25-30. Double leaf bracelet $35-45. Earrings $15-20.

Unusual reversible bracelet. Plastic stones are hinged and can be changed from white to pink. c.1960s. Unsigned. $35-40.

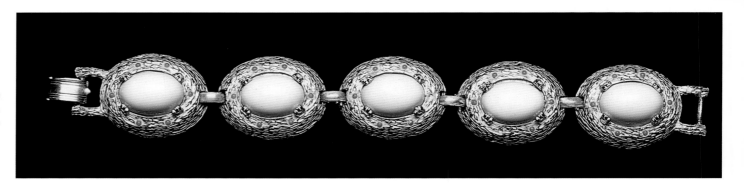

This bracelet looks like it has little white eggs, sitting atop nests. c.1960s. Unsigned. $30-35.

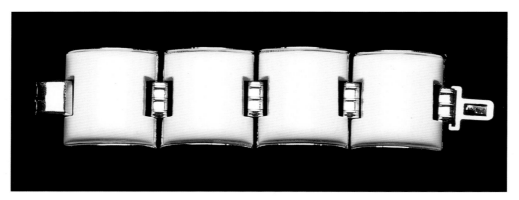

Large, white mod bracelet. c. 1960s. Unsigned. $30-35.

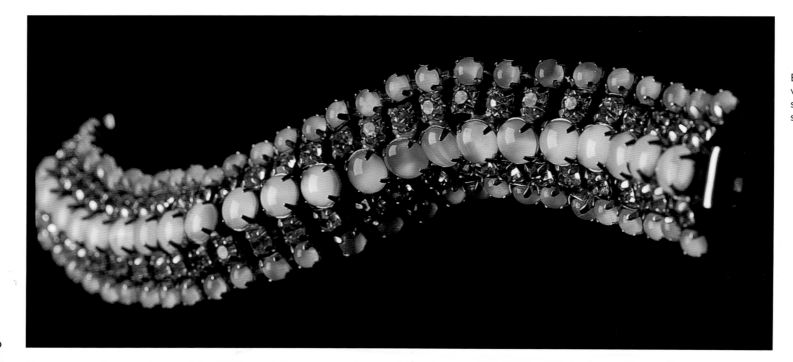

Elegant Warner bracelet with opaque-white plastic stones and clear rhinestones. c.1950s. $40-50.

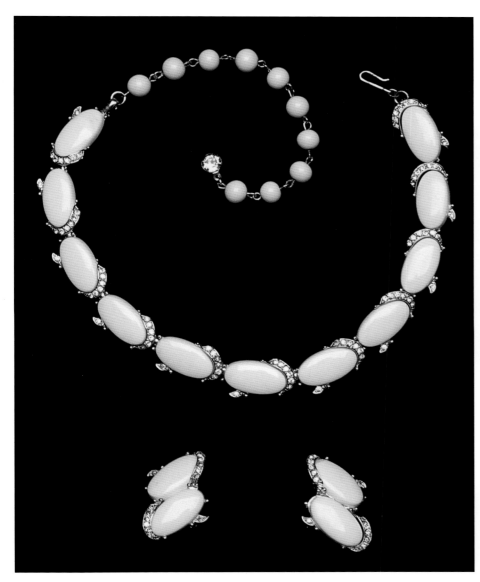

Necklace and earrings of white ovals ringed in rhinestones. c.1950s. Signed Kramer. Jewelry courtesy of Cheryl Killmer. Necklace $30-35. Earrings $15-20.

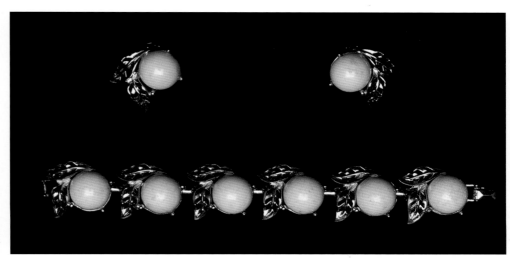

Big white dots are featured in this bracelet and earrings set. c.1950s. Unsigned. Jewelry courtesy of Cheryl Killmer. Bracelet $25-30. Earrings $10-15.

Delicate, abstract floral design in bracelet and earrings. c.1950s. Signed Trifari. Jewelry courtesy of Cheryl Killmer. Bracelet $45-50. Earrings $20-25.

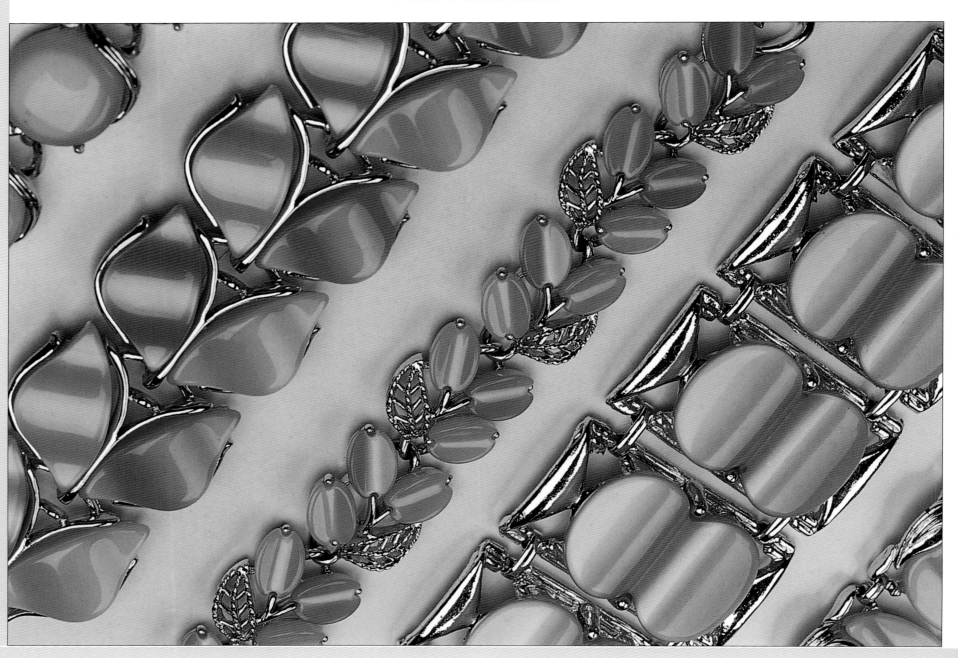

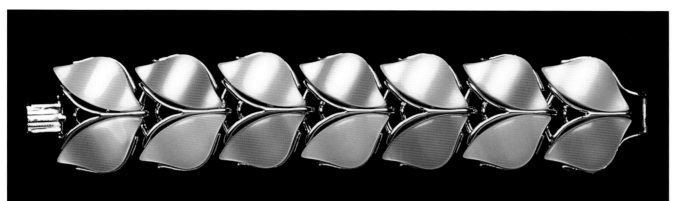

Large, two-tone pink leaves bracelet.
c.1950s. Unsigned. $40-50.

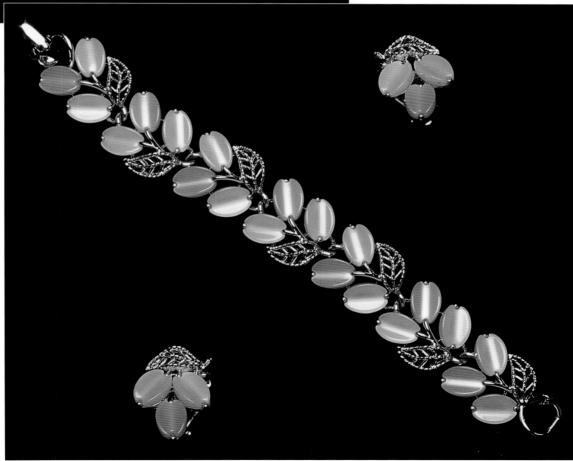

Pink tulip-like bracelet and earrings. c.1950s.
Lisner. Bracelet $40-45. Earrings $15-20.

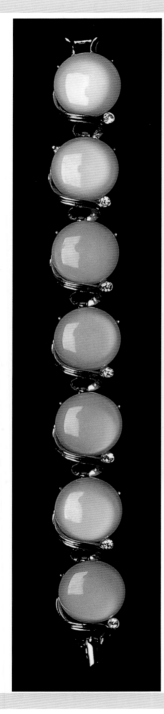

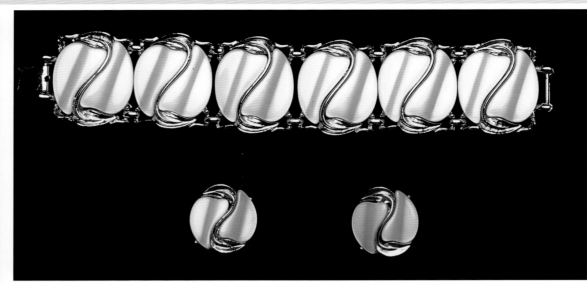

Findings of gold vines bisect this big, pink beauty. c.1950s. Unsigned. Bracelet $35-40. Earrings $15-20.

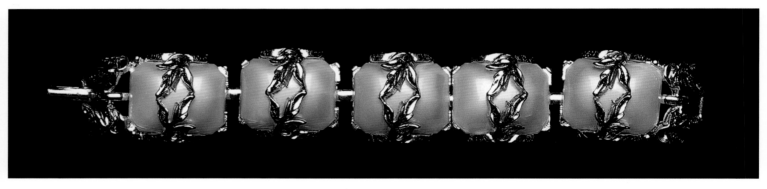

Silver findings wrap around pink, curved oblongs. c.1950s. Unsigned. $30-35.

This is the jewelry featured in the 1957 Dayton's Department Store Lisner ad "Lisner's pretend jewels glow where you go" (Page 46). The bracelet originally sold for $1, plus tax! Today $30-35. Not a bad investment!

Right:
Ladylike, iridescent leaves and pearls in a bracelet and earrings set. c.1950s. Unsigned. Bracelet $35-40. Earrings $20-25.

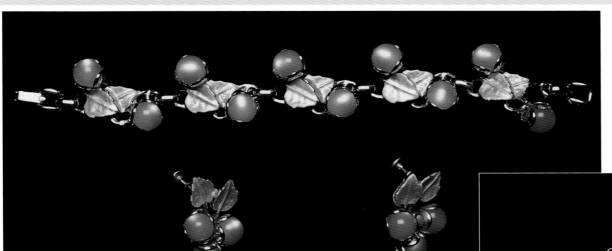

Moonglow-type spheres, set among dainty gold leaves in a bracelet and screw-back style earrings set. c. 1950s. Unsigned. Bracelet $35-40. Earrings $15-20.

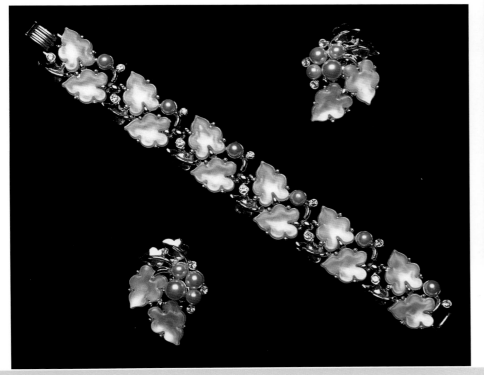

Similar necklace and bracelet also utilizes pink moonglow-type spheres. This set backed with gold disks. c. 1950s. Unsigned. Necklace $35-40. Bracelet $30-35.

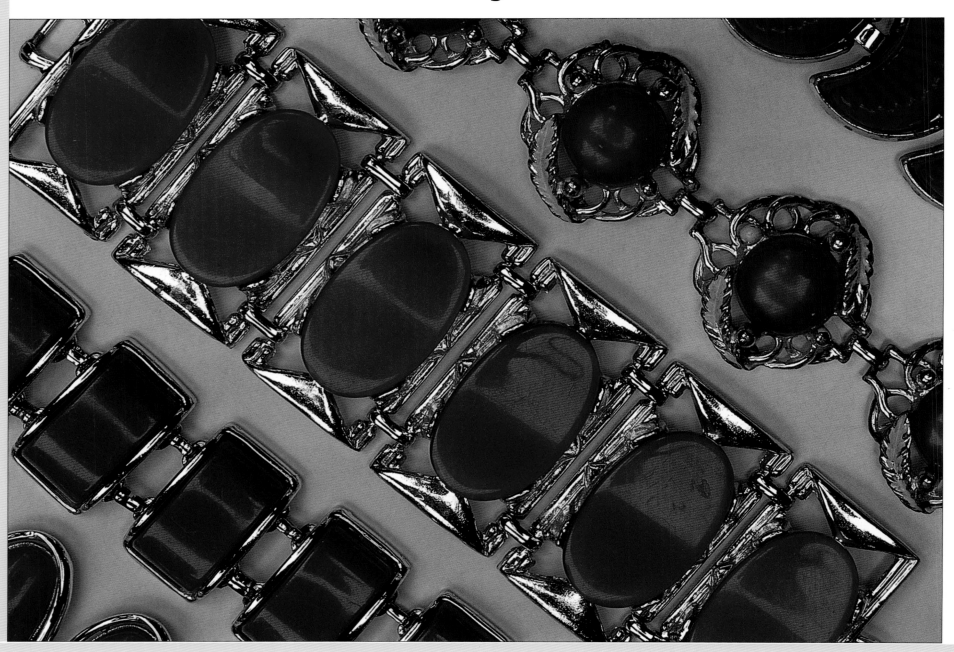

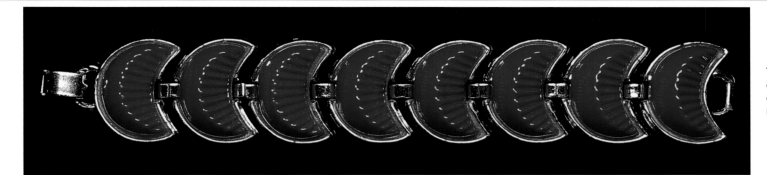

This bracelet looks like it has delectable sections of citrus. c. 1960s. Unsigned, but tasty nonetheless. $35-40.

Clean lines illustrate this large, cranberry red bracelet. c. 1950s. Unsigned. Jewelry courtesy of Cheryl Killmer. $45-50.

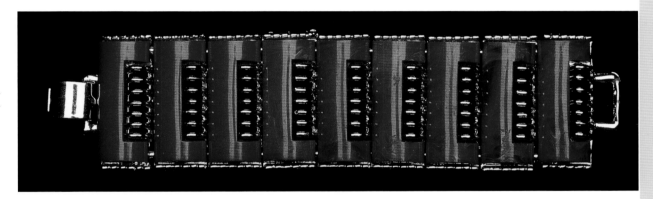

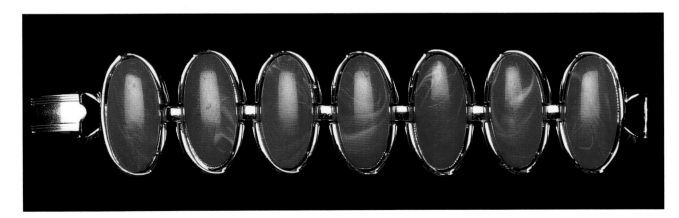

Bracelet of long red ovals that are swirled with white. c. 1950s. Unsigned. $30-35.

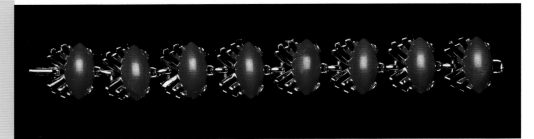

The gold tone findings on this red bracelet make it look like a line of crawling crabs. c.1950s. Unsigned. $30-35.

The plastic parts on this bracelet are joined by an "S" and since I'm "Sue" I had to buy it! c.1950s. Unsigned. $35-40.

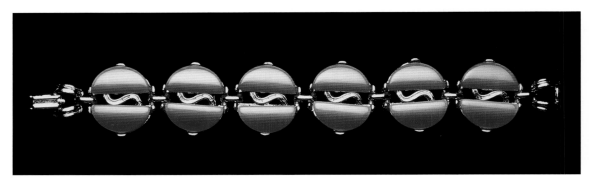

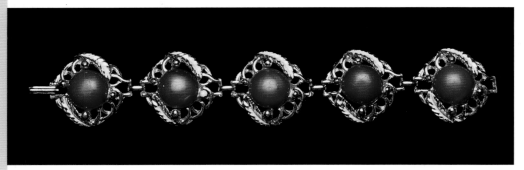

Looking like juicy, red cherries, these plastic stones are enveloped by white painted-over gold tone findings. c.1950s. Unsigned. $30-35.

Satinore-type stones set into findings that are encrusted in rhinestones. c.1950s. Unsigned. $40-50.

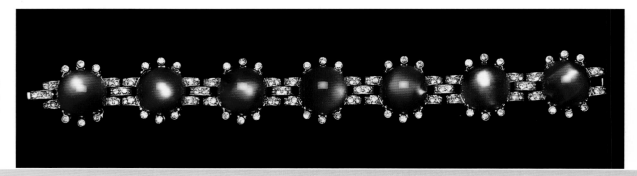

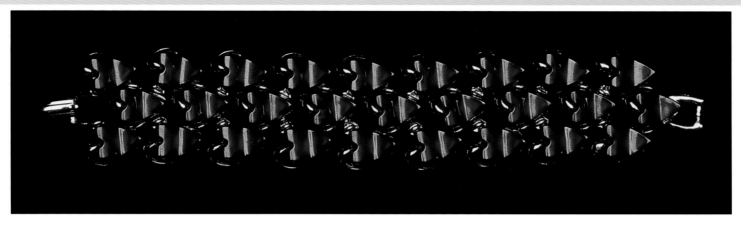

Large bracelet of deep red, wavy hearts. c. 1960s. Unsigned. Jewelry courtesy of Judy Levin. $45-55.

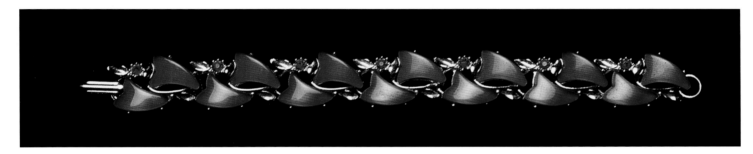

The little, red arrows point to an attractive bracelet. c.1950s. Unsigned. $30-35.

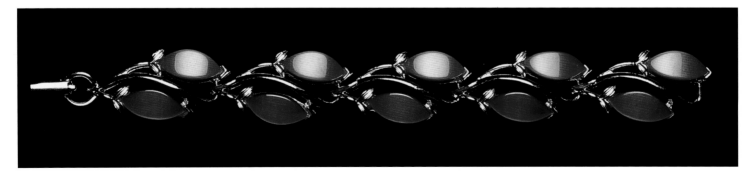

Silver tendrils bloom with red plastic stones. c.1950s. Unsigned. $30-35.

Flutter by this lovely set. c.1950s. Unsigned. Necklace $30-35. Earrings $15-20.

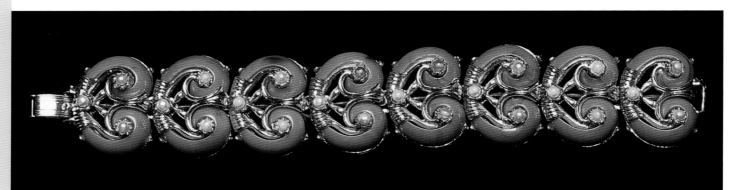

Jack-o-lantern-like bracelet. c. late 1940s-early 1950s. Unsigned. $40-50.

Yin and yang. Translucent pieces in orangey-amber are probably made of polystyrene. The hardness of the plastic is the tip-off. c. 1950s. Unsigned. $35-40.

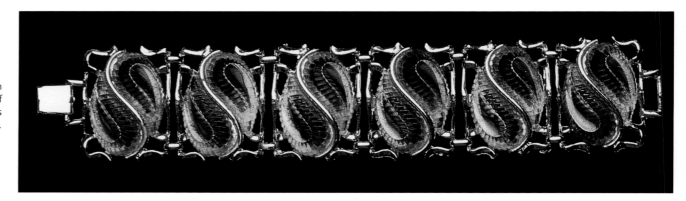

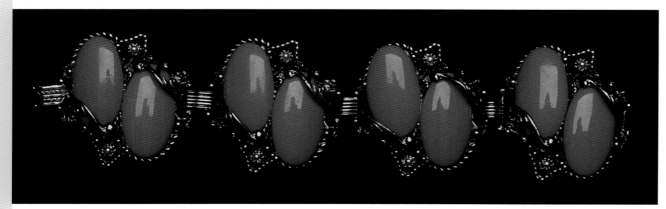

Pairs of deep coral ovals emphasize this bracelet. c.1950s. Unsigned. Jewelry courtesy of Cheryl Killmer. $40-50.

Orange disks delightfully dangle from this charm bracelet. Carol Sullivan wore it to the photo shoot, and it was so fabulous that we made her take it off so we could photograph it for the book! c. late 1940s. Unsigned. Jewelry courtesy of Carol Sullivan. $40-50.

This Coro bracelet features teardrop shapes cradled in a gold tone setting. c.1950s. $35-40.

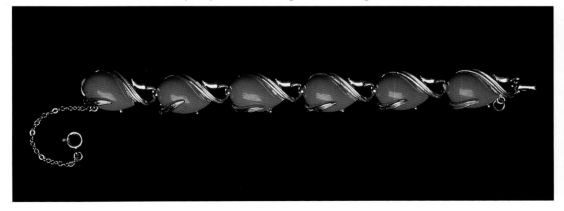

One of my personal favorites. Coral cabochons are set atop a weighty gold tone setting. c. mid-1960s. Unsigned. $40-50.

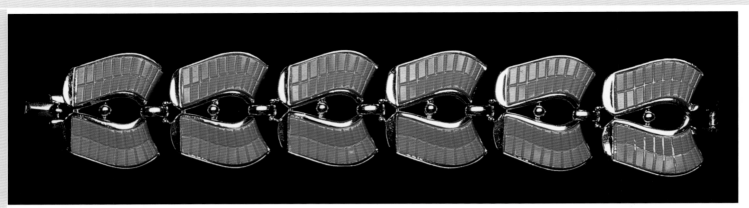

Uniquely shaped, texturized orange bracelet. c. 1960s. Signed H & S. $35-40.

Charm bracelet of pear-shaped, orange polystyrene drops, swirled with black. c. late 1940s-early 1950s. Unsigned. $30-35.

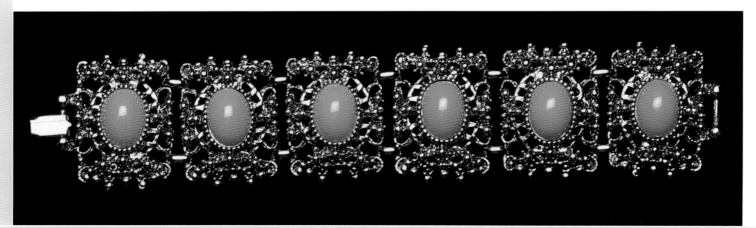

This bracelet's elaborate setting is topped with orange ovals. Note the green bracelet on page 18; it has the same hardware, but a very different plastic insert. c. 1960s. Unsigned. $30-35.

Graceful bracelet and earrings with teardrop shapes in orange and yellow. c.1950s. Unsigned. Bracelet $35-40. Earrings $20-25.

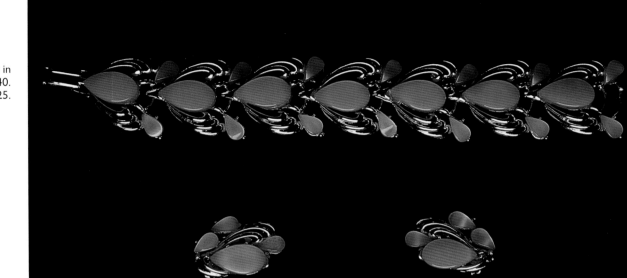

This Coro set is hot Stuff! Chili pepper-like earrings and matching bracelet. c.1950s. Bracelet $40-45. Earrings $20-25.

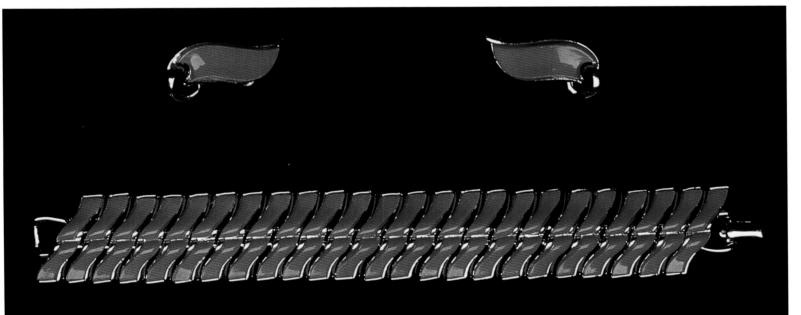

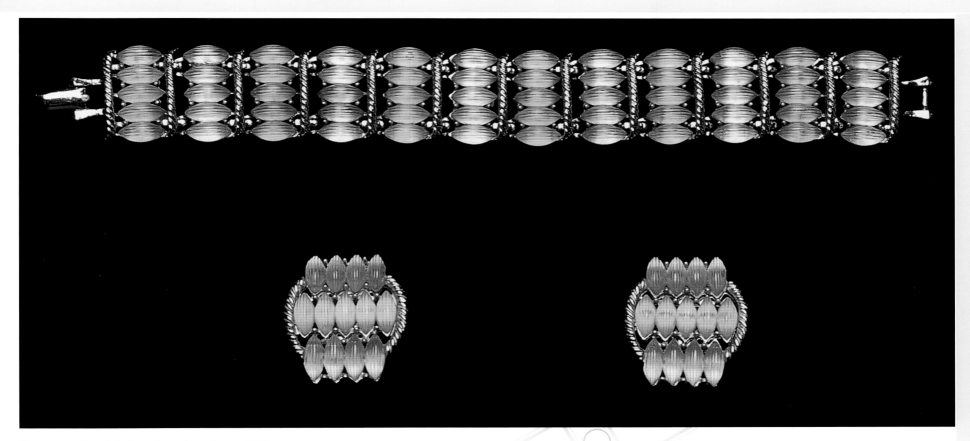

Two-tone yellow, jelly bracelet and earrings. c.1950s.
Unsigned. Bracelet $40-45. Earrings $20-25.

Floral-detailed findings combine with
banana-shaped parts in this bracelet.
c.1950s. Unsigned. $30-35.

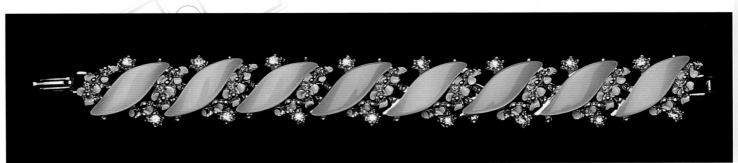

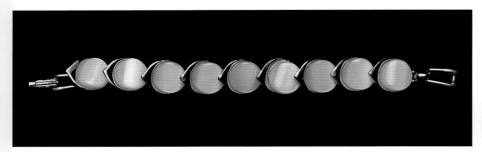

Bracelet of yellow disks and pointy hardware. c.1950s. Unsigned. $30-35.

These Lisner knots won't be forgotten. c. 1960s. $35-40.

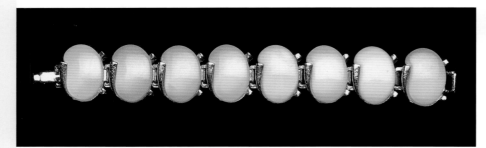

Lemon-yellow ovals by Leru. c.1950s. $30-35.

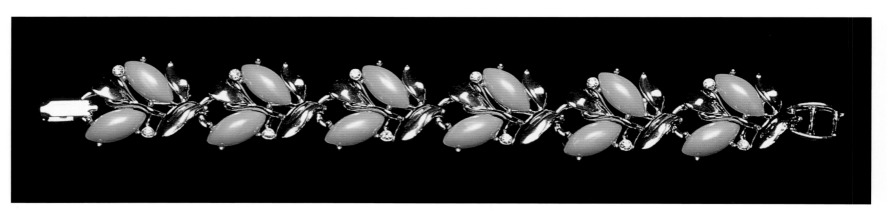

Delicate yellow bracelet with a floral design. c.1950s. Unsigned. $30-35.

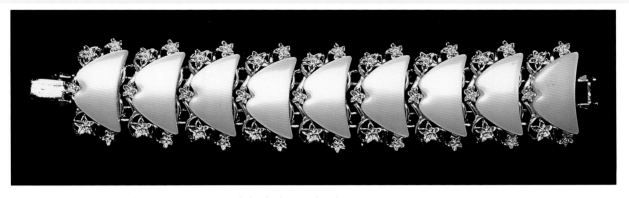

These plastic parts look like mini mountains and the findings utilized
in this bracelet are extremely heavy. c.1950s. Unsigned. $40-50.

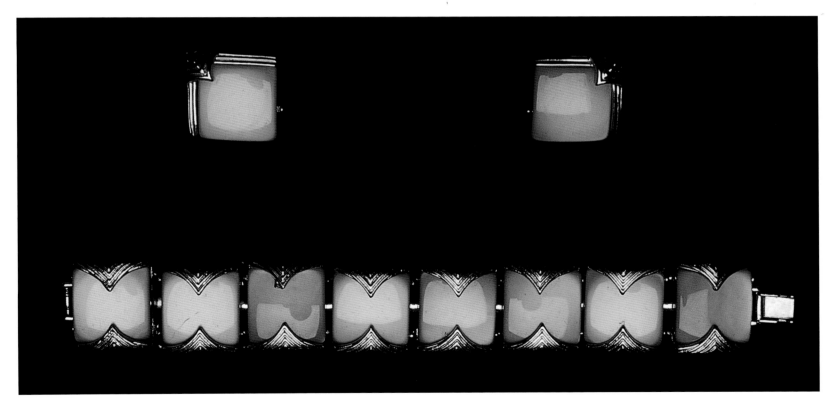

Bracelet and earrings of yellow squares sandwiched between coppery findings. c. 1950s.
Signed Coro. Jewelry courtesy of Judy Levin. Bracelet $30-35. Earrings $12-15.

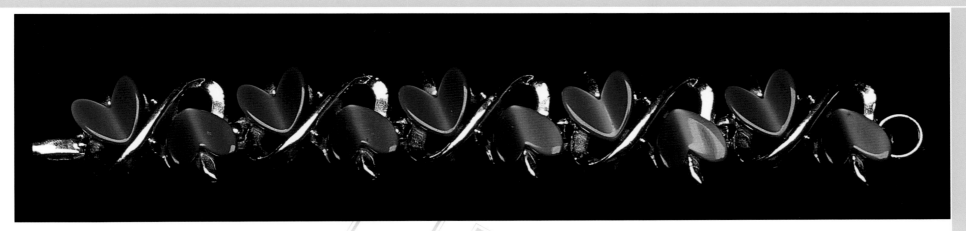

You can wear your heart on your sleeve, or even on your wrist!
Green heart bracelet. c.1950s. Unmarked. $30-35.

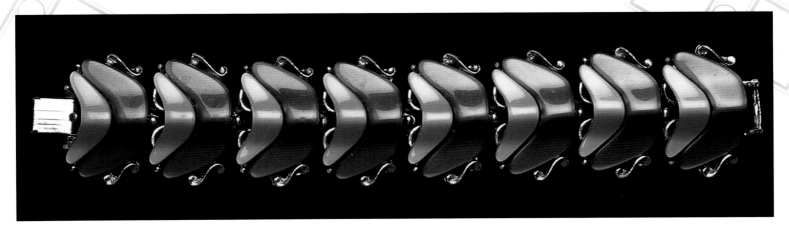

Two-tone green bracelet with free-form plastic shapes. c.1960s. Unmarked. $35-40.

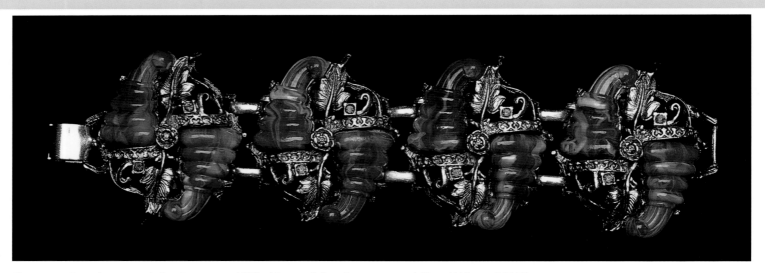

Cornucopia bracelet in mottled, jade green. c.1950s. Unsigned. Jewelry courtesy of Cheryl Killmer. $45-55.

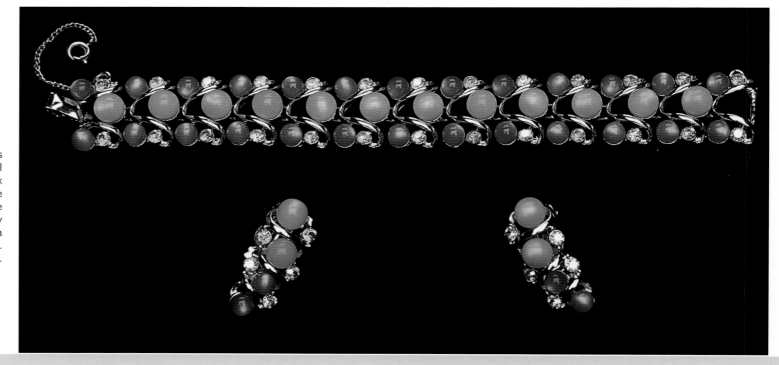

Magnificent yellow-green spheres in a bracelet and earrings by Star. I found the earrings first, then six months later I matched-up the bracelet! When trying to complete sets – keep looking! Eventually you'll find it, and, obviously, a complete set adds value. c.1950s. Bracelet $35-40. Earrings $15-20.

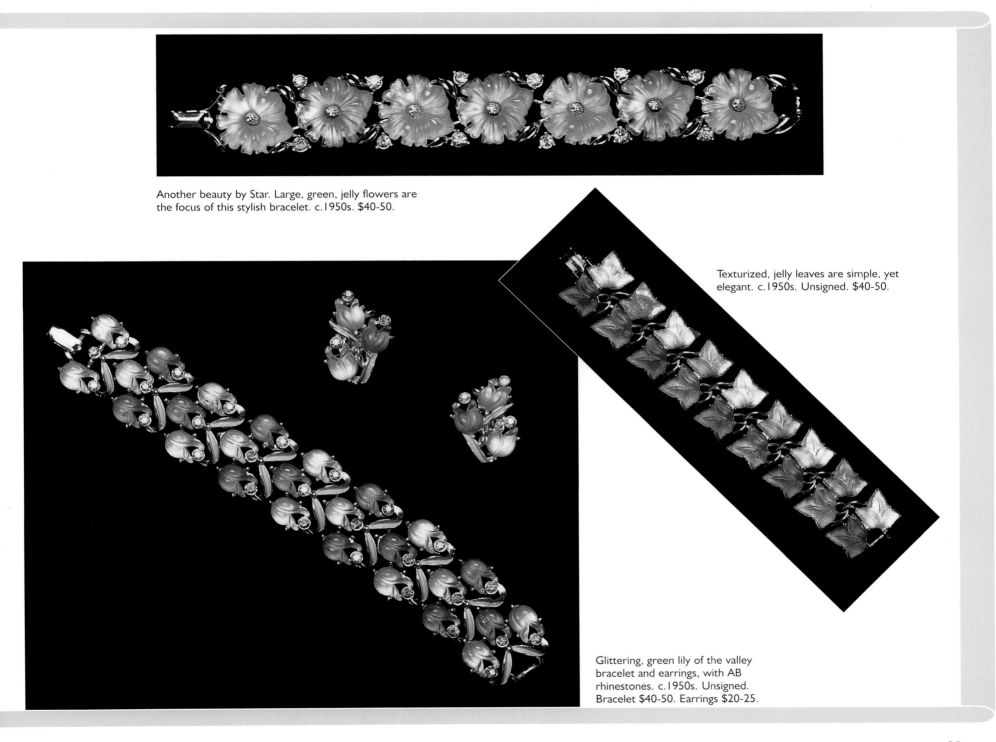

Another beauty by Star. Large, green, jelly flowers are the focus of this stylish bracelet. c.1950s. $40-50.

Texturized, jelly leaves are simple, yet elegant. c.1950s. Unsigned. $40-50.

Glittering, green lily of the valley bracelet and earrings, with AB rhinestones. c.1950s. Unsigned. Bracelet $40-50. Earrings $20-25.

89

Leaf bracelet in vivid shades of green. The leaf's stem is actually crafted into the findings of the bracelet. c.1950s. Unsigned. $35-40.

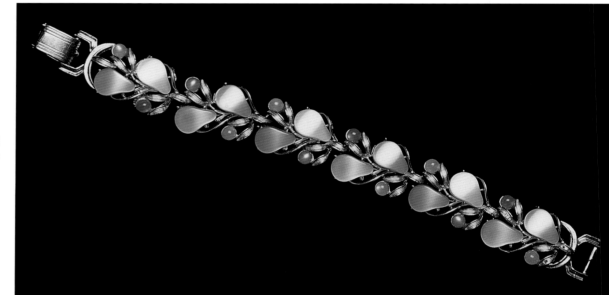

Coro bracelet with green, pear-shaped ornamentation. c.1950s. $30-35.

Pearls combine with irregularly-shaped green stones in this bracelet. c.1950s. Unsigned. Jewelry courtesy of Cheryl Killmer. $30-35.

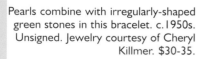

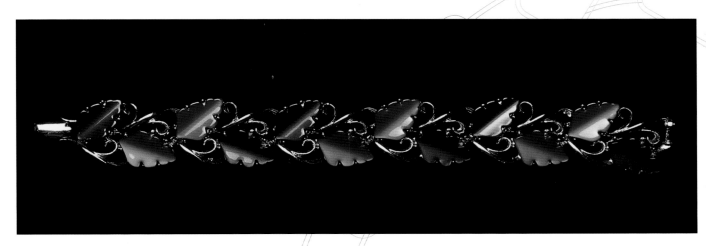

A Lisner leaf knock-off. This graceful, leaf bracelet is
unsigned. c.1950s. $30-35.

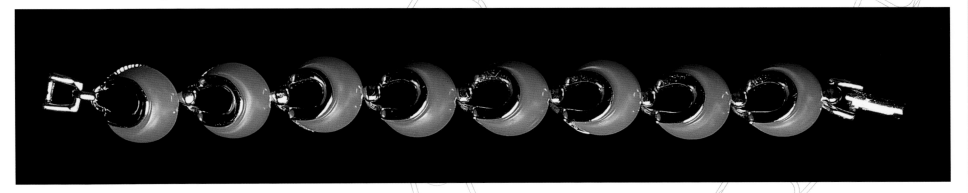

Bracelet of green-blue crescents set into gold findings.
c.1960s. Unsigned, but possibly Karu. $30-35.

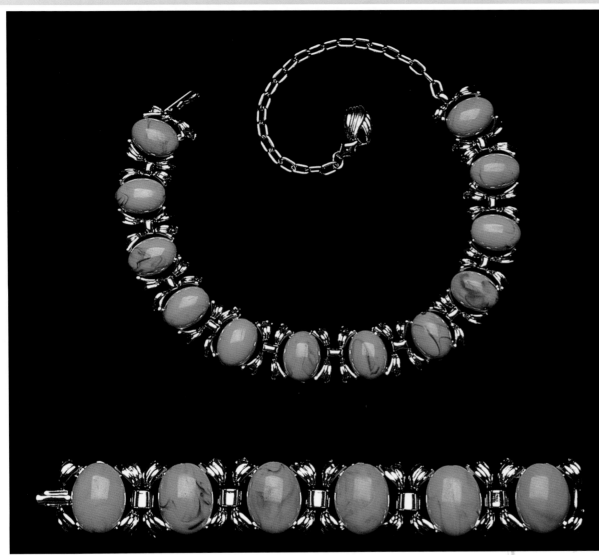

Faux turquoise necklace and bracelet. c.1960s. Unsigned. Necklace $30-35. Bracelet $25-30.

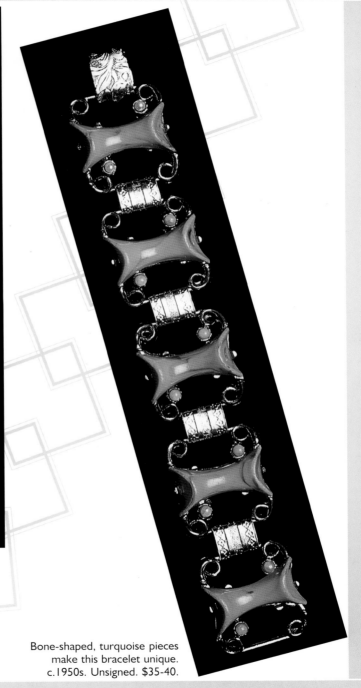

Bone-shaped, turquoise pieces
make this bracelet unique.
c.1950s. Unsigned. $35-40.

"X" marks the spot on this turquoise, plastic bracelet by Kramer. c. 1950s. $30-35.

The individual links on this monster, turquoise bracelet are a full 1.5" wide! The earrings measure 1.0" c.1950s. Unsigned. Bracelet $40-50. Earrings $20-25.

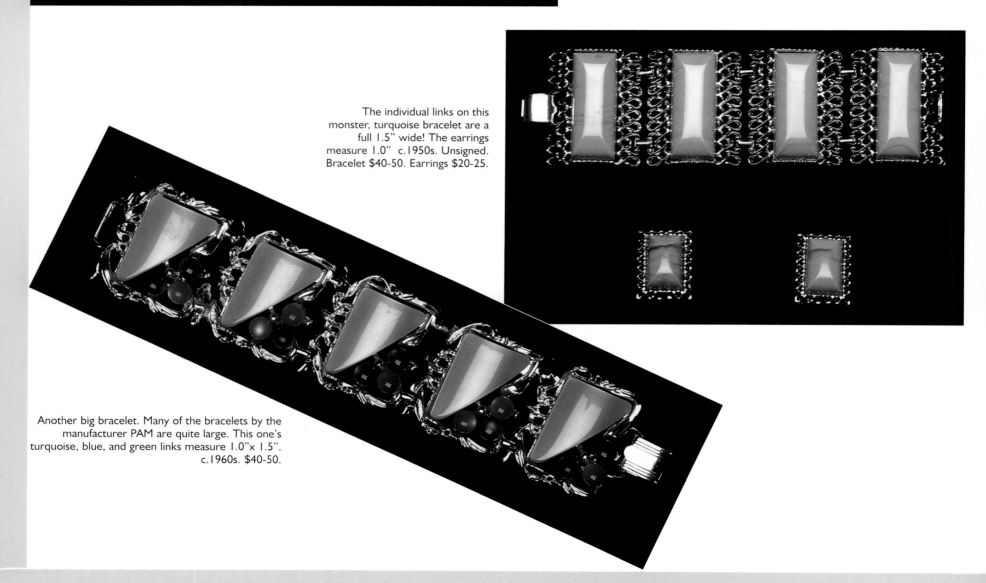

Another big bracelet. Many of the bracelets by the manufacturer PAM are quite large. This one's turquoise, blue, and green links measure 1.0"x 1.5". c.1960s. $40-50.

Enormous pendant-style necklace and bracelet featuring disks of mock turquoise. This set has extraordinarily heavy findings. c.1950s. Unsigned. Necklace courtesy of Cheryl Killmer. Bracelet belongs to the author. Necklace $35-40. Bracelet $35-40.

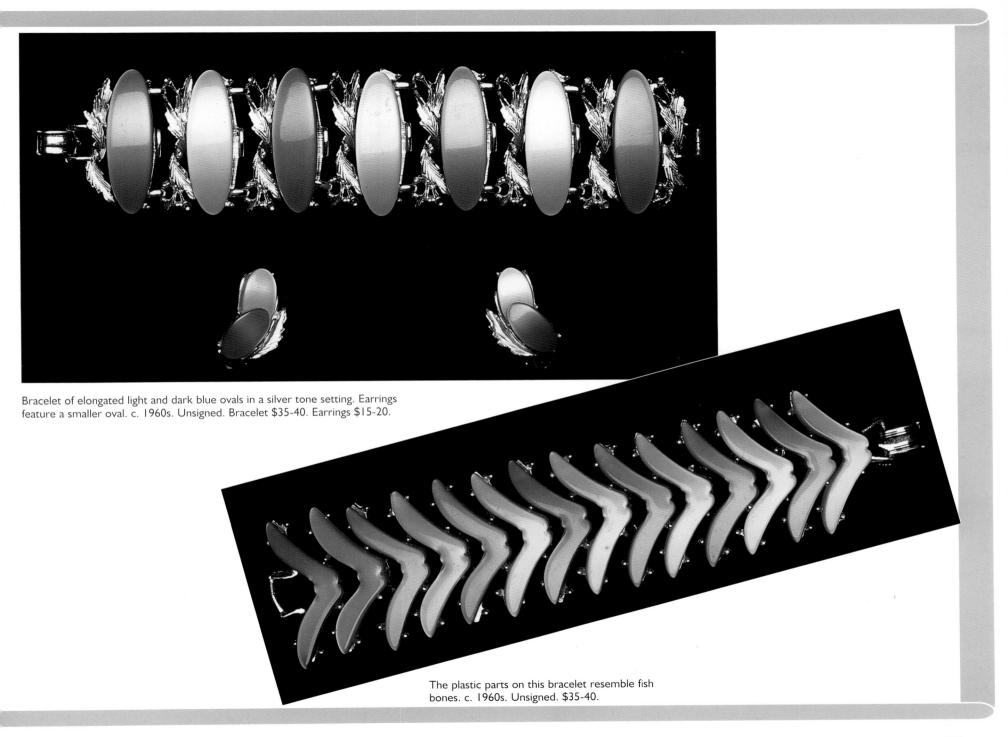

Bracelet of elongated light and dark blue ovals in a silver tone setting. Earrings feature a smaller oval. c. 1960s. Unsigned. Bracelet $35-40. Earrings $15-20.

The plastic parts on this bracelet resemble fish bones. c. 1960s. Unsigned. $35-40.

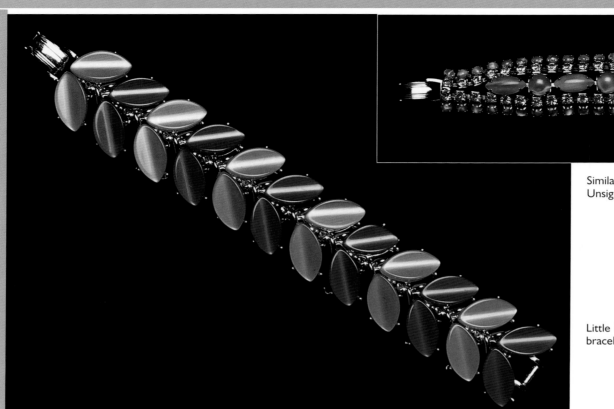

Similar bracelet to the one below, with pearly blue plastic stones. Unsigned. $40-50.

Little leaves comprise this Kramer bracelet. c. 1960s. $30-35.

Simple bracelet of blue plastic oblongs and rhinestones. The plastic used truly resembles glass. Earrings to match. c.1950s. Unsigned. Bracelet $35-40. Earrings $20-25.

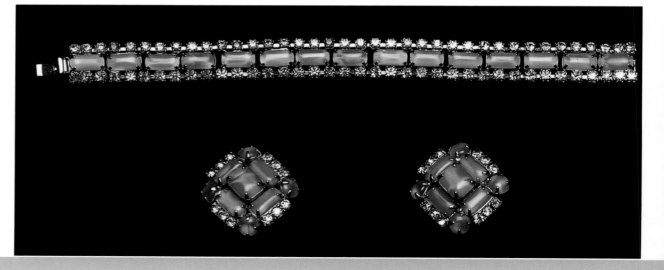

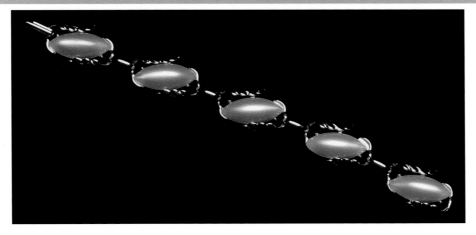

Light blue, opaline cat-eye shapes set into silver tone plated, wings that are linked to make this bracelet. c.1950s. Unsigned. $30-35.

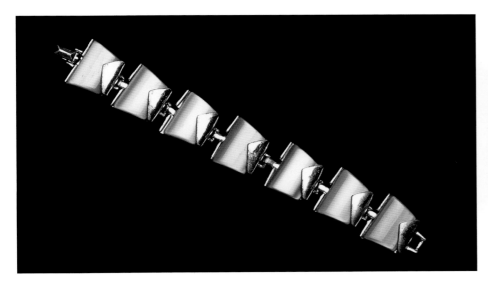

This bracelet's parts look like miniature purses! c.1950s. Charel. $35-40.

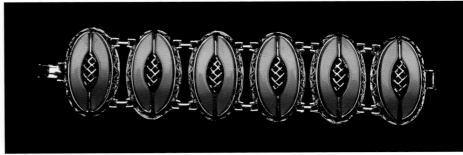

Gold-tone filigree findings completed with large, aqua blue ovals. c.1960s. Unsigned. $40-50.

Bracelet of iridescent blue cabochons. The plastic used in this piece is characteristic of dipped polystyrene. c.1960s. Kramer. $30-35.

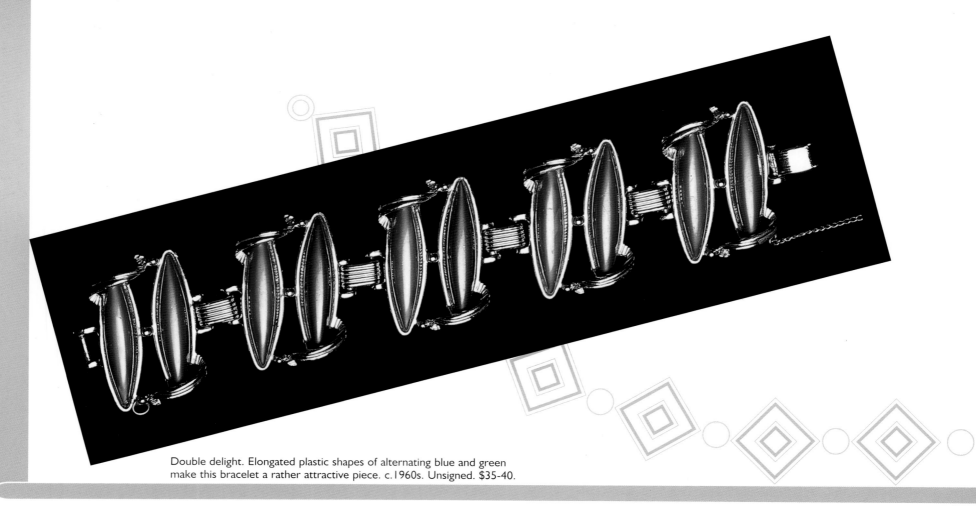

Double delight. Elongated plastic shapes of alternating blue and green make this bracelet a rather attractive piece. c.1960s. Unsigned. $35-40.

Blueberry colored bracelet and earrings look good enough to eat! c.1950s. Unsigned. Bracelet $25-30. Earrings $10-15.

Praise Purple

A Coro copy. These lilac squares are unsigned. c.1950s. $30-35.

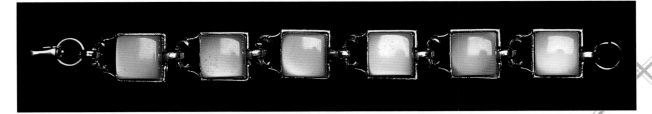

Bracelet of lavender, candle shaped parts. c. 1950s. Unsigned. $35-40.

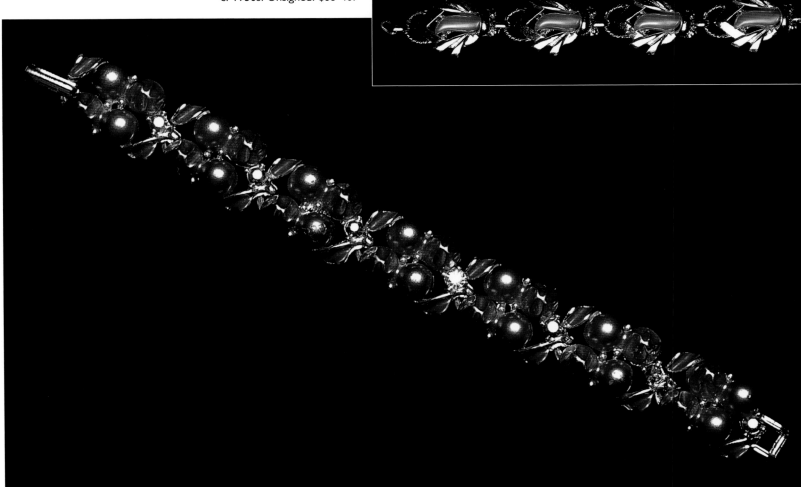

Purple pearls and AB rhinestones. c. 1950s. Unsigned. $35-40.

Orchid plastic, petals blossom on this bracelet. c.1950s. Unsigned. $35-40.

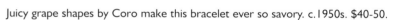

Juicy grape shapes by Coro make this bracelet ever so savory. c.1950s. $40-50.

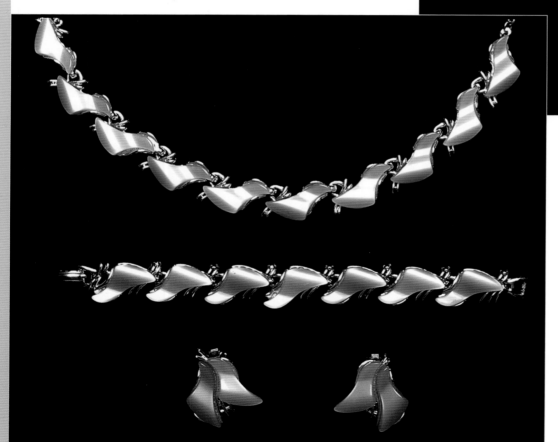

Pink and purple parure of necklace, bracelet, and earrings by Art. Very 1950s! Set $75-100.

Right:
The perfect necklace, bracelet, and earrings to wear with your poodle skirt. Pink and purple parure. c. 1950s, of course! Set $75-100.

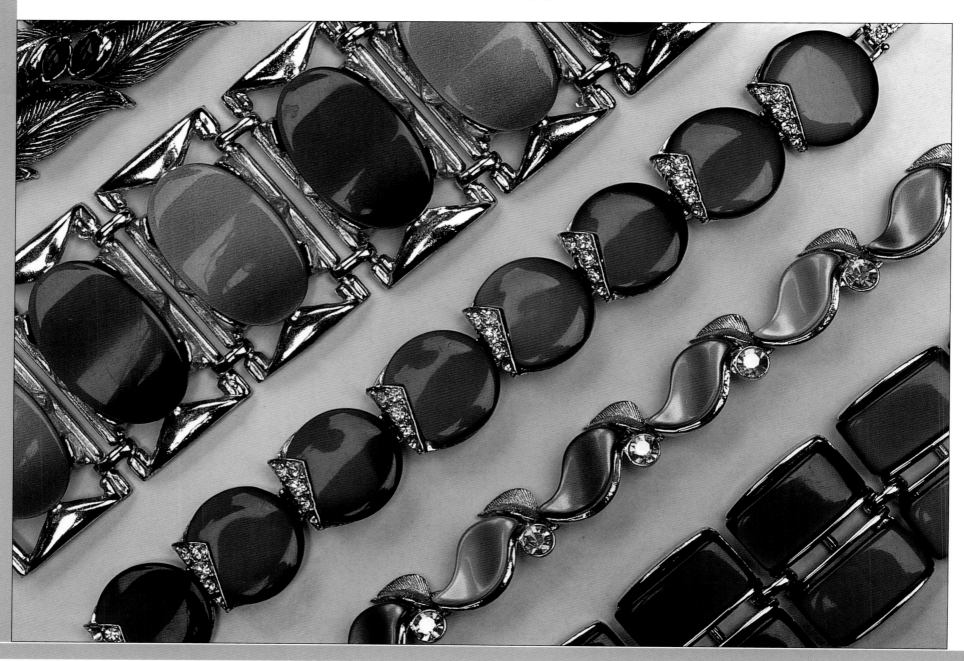

Pearly, butterscotch and brown dots in a simple, but attractive bracelet. Unsigned. c.1960s. $30-35.

Brown circles encircle this bracelet. Unsigned. c.1960s. $30-35.

Bracelet of golden brown squiggles that has the characteristic look of dipped polystyrene. Unsigned. c.1960s. $30-35.

Florenza (Larry Kassoff) bracelet of graceful metal leaves topped with small, brown plastic nuggets. c.1960s. $40-50.

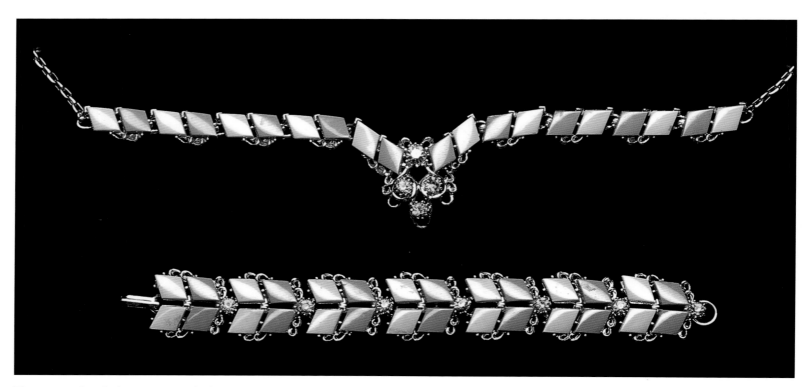

The toasty colored, plastic pieces in this bracelet and necklace set are quite thin, which is characteristic of cellulose acetate. Unsigned. c. 1950s. Necklace $35-40. Bracelet $30-35.

Trio of large, animal print bracelets. Plastic material used is thin and translucent and is, therefore, indicative of cellulose acetate. Center bracelet appears to have been blow-molded. (See page 14 for description of blow-molding.) (The center bracelet is also a larger version of the citrus red bracelet on page 73.) All unsigned. c.1960s. Top and bottom bracelet courtesy of Cheryl Killmer. Center bracelet belongs to the author. Each bracelet $40-50.

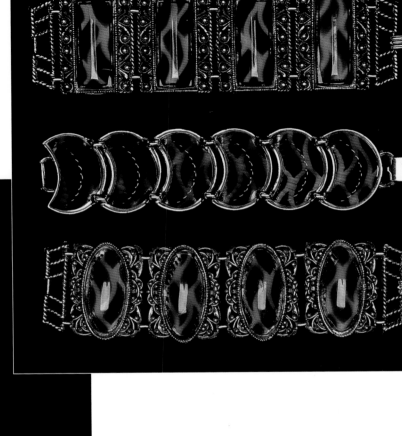

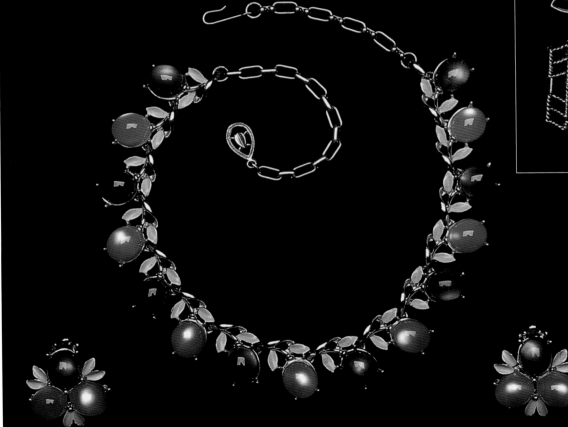

Satiny brown and tan necklace and earrings. Unsigned. Jewelry courtesy of Cheryl Killmer. Necklace $45-55. Earrings $20-25.

Bold Beige

Glowing, creamy orbs are featured in this Coro Set.
c.1950s. Bracelet $30-35. Earrings $15-20.

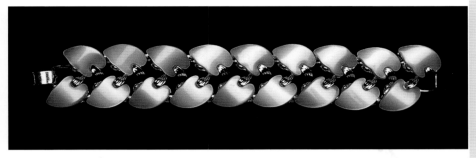

Tawny, moderne leaves encompass this bracelet. c.1950s.
Unsigned. Jewelry courtesy of Cheryl Killmer. $35-45.

The tiny leaves in this bracelet actually have several, slight color variations which give it much depth. c.1950s. Notice this bracelet's nearly identical twin on page 54, which is marked Kramer. This piece is unsigned. $35-40.

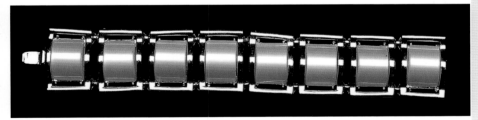

Bracelet of pearly off-white segments that float in gold-tone rectangles. c.1960s. Unsigned. $30-35.

This Claudette bracelet looks like a bunch of bananas! c. 1950s. $35-40.

Little raindrop shaped pieces between rows of AB rhinestones. c.1950s. It seems too attractive a bracelet to actually be unsigned! $40-50.

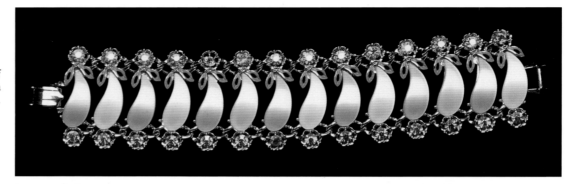

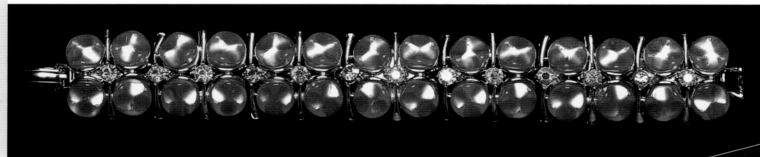

A most unique piece. Can you tell what the plastic parts really are?

Answer: They're buttons! I discovered this when one of the pieces popped out and to my surprise it was a real button. c.1950s. Signed Art. $30-35.

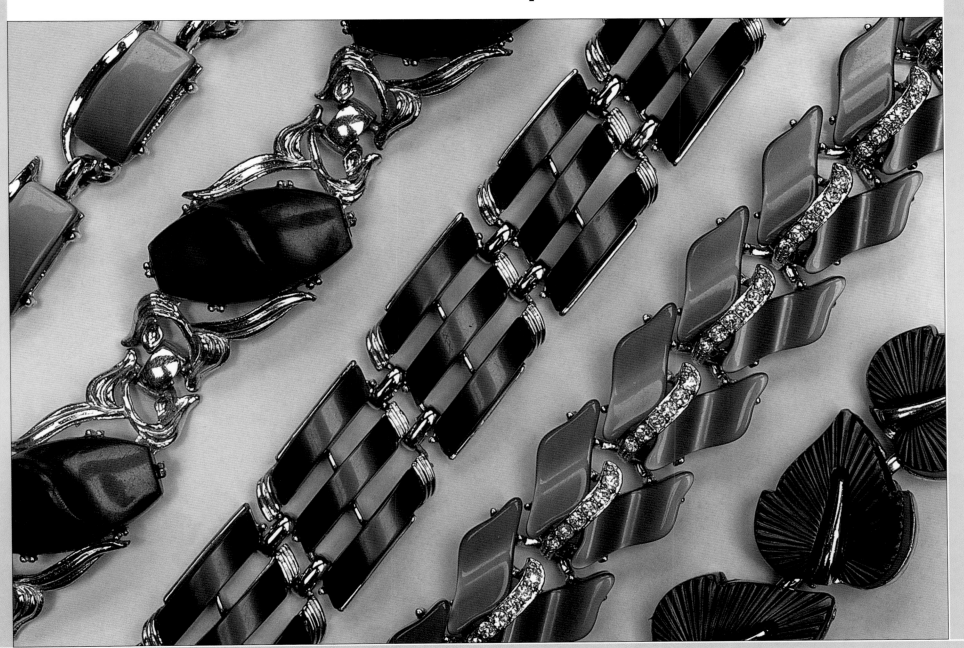

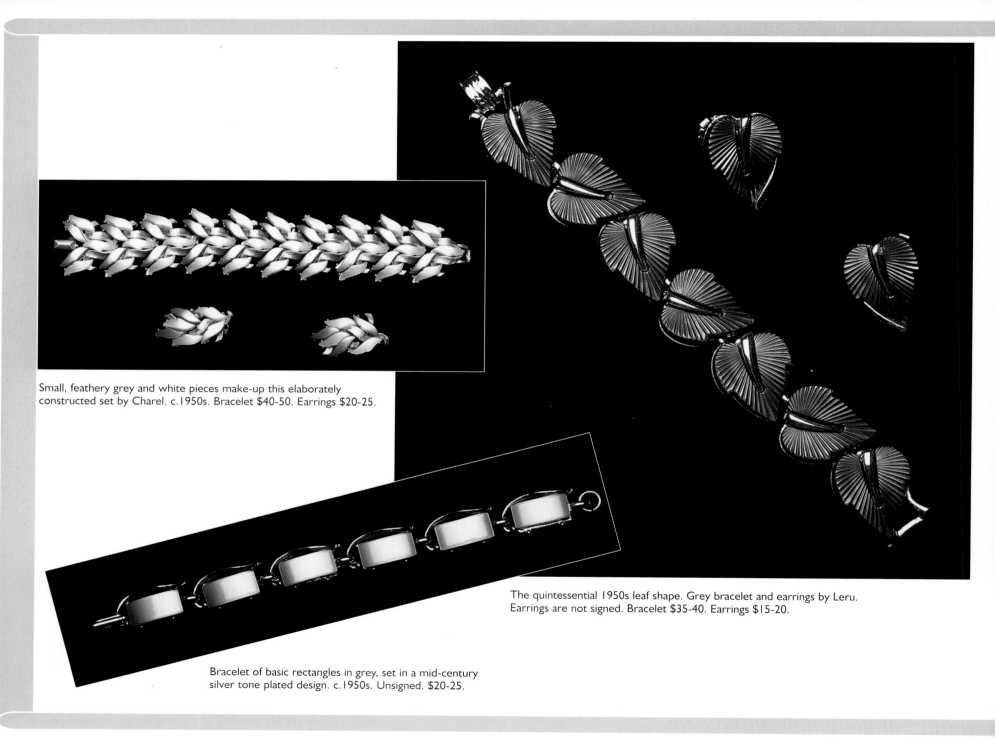

Small, feathery grey and white pieces make-up this elaborately constructed set by Charel. c.1950s. Bracelet $40-50. Earrings $20-25.

The quintessential 1950s leaf shape. Grey bracelet and earrings by Leru. Earrings are not signed. Bracelet $35-40. Earrings $15-20.

Bracelet of basic rectangles in grey, set in a mid-century silver tone plated design. c.1950s. Unsigned. $20-25.

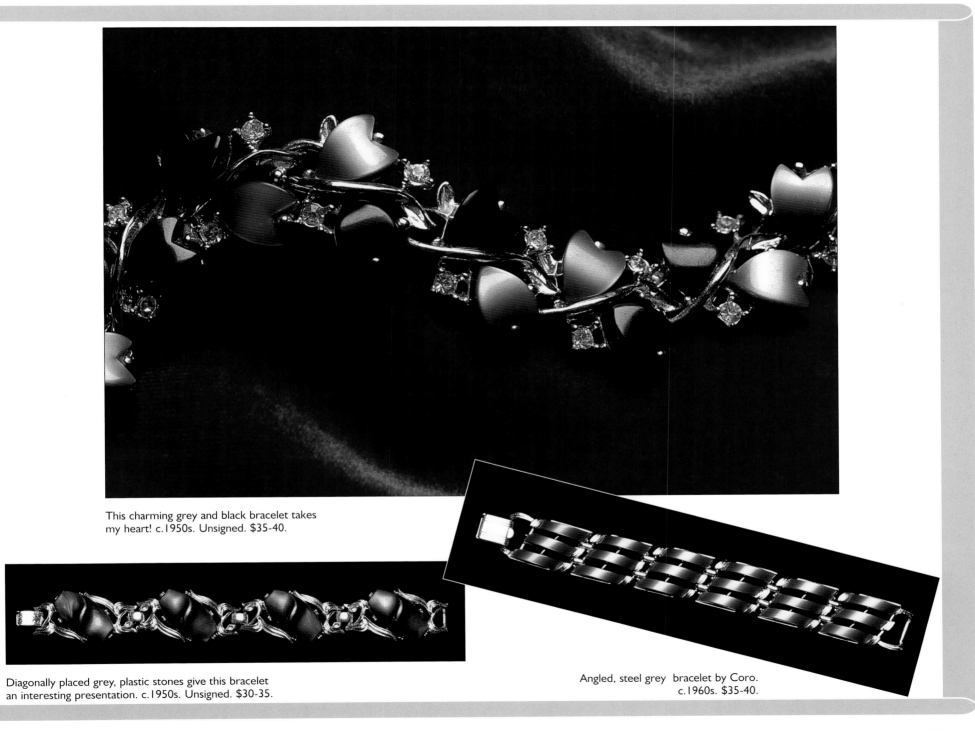

This charming grey and black bracelet takes
my heart! c.1950s. Unsigned. $35-40.

Diagonally placed grey, plastic stones give this bracelet
an interesting presentation. c.1950s. Unsigned. $30-35.

Angled, steel grey bracelet by Coro.
c.1960s. $35-40.

Charm bracelet of large gold-tone leaves and black plastic dangles. Unsigned. c.1950s. $30-35.

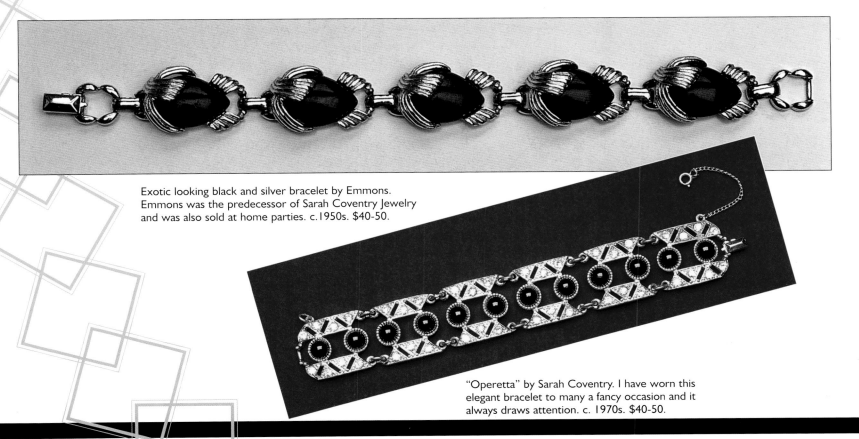

Exotic looking black and silver bracelet by Emmons. Emmons was the predecessor of Sarah Coventry Jewelry and was also sold at home parties. c.1950s. $40-50.

"Operetta" by Sarah Coventry. I have worn this elegant bracelet to many a fancy occasion and it always draws attention. c. 1970s. $40-50.

Black and white and white and black. From top to bottom: Bracelet of slanted ovals. Unsigned. c.1960s. $30-35. Bracelet of tiny, alternating ovals. Unsigned. c. 1950s. $35-40. Matching earrings are at bottom of shot $20-25. Bracelet of black and white triangles. Unsigned. c.1960s. $30-35.

Large black ovals on bracelet, small ones on earrings. Unsigned. c.1950s. Bracelet $30-35. Earrings $15-20.

Multiples

A pleasure of collecting is finding the same design in a multitude of colors and variations.

Delightful daisies Top lavender and violet, bottom green and blue. Unsigned. c.1960s. Each $40-50.

Same findings, yet different bracelets. On the left, dark green plastic baubles and rhinestones. On the right, a light green variation. c.1950s. Unsigned. $40-50, each. The hand beaded cashmere sweater belonged to my great Aunt Rose – a woman who always wore fabulous costume jewelry and, best of all, let me raid her jewelry box!

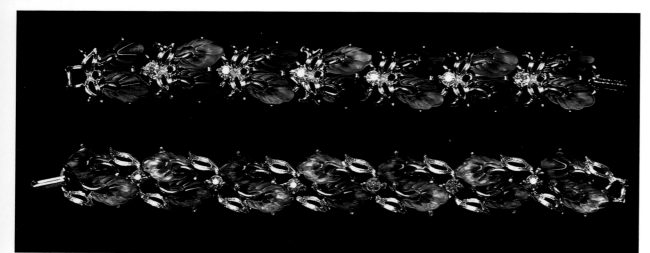

Texturized jelly leaves. Top bracelet in two shades of green, bottom bracelet in two shades of blue. Lisner look-alikes, but this pair is unsigned. c.1960s. Each $35-40.

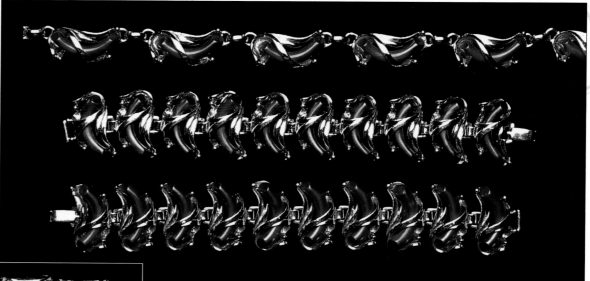

I refer to this grouping as the "hot dog" set. Fantastically shaped necklace and bracelet in cherry red and additional bracelet in royal blue. Unsigned. c.1950s. Necklace $35-40. Bracelets each $40-50.

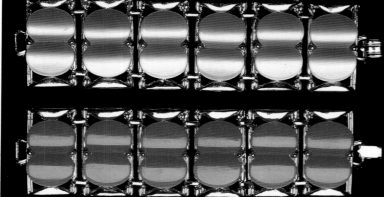

Two more very mod PAM bracelets, one light pink, one hot pink. c.1960s. Each $40-50.

It looks like a pop art painting, but it's really four bracelets by PAM. The variations are day-glo pink and orange, tan and brown, red, and deep purple and violet. c.1960s. Each $40-50.

Split circle parure in lavender and white. Unsigned. c.1960s. Jewelry and photo courtesy of The Francesca Medrano Collection. Set $75-95.

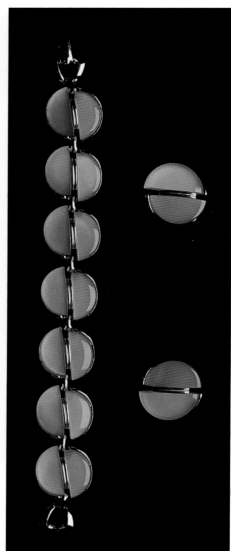

Yellow and white version of the split circle set. Unsigned. c.1960s. Bracelet $40-50. Earrings $20-25.

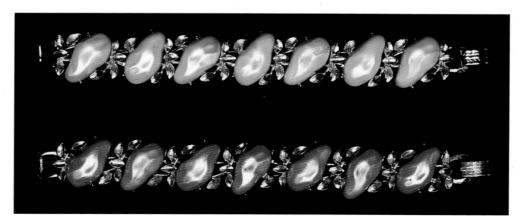

Pear-shaped yellow and coral bracelets by B.S.K. (The B. Steinberg-Kaslo Co. was composed of partners Julius Steinberg, Morris Kimmelman, Hyman Slovitt, Abraham J. Slovitt, Samuel Friedman, and another entity called Arke, Inc. This information was gleaned from the text of a 1956 copyright infringement lawsuit between Trifari and B.S.K.) c. 1950s. Each $40-50.

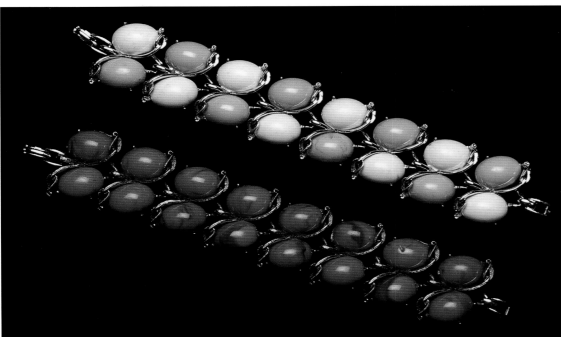

Two lovely, linked bracelets. Top bracelet is a two-tone tan, while bottom bracelet is a mottled black and orange. Unsigned. c.1950s. Each $40-50.

Whimsical under-the-sea charm bracelets accented with orange plastic drops. Unsigned. c.1950s. Each $35-40.

123

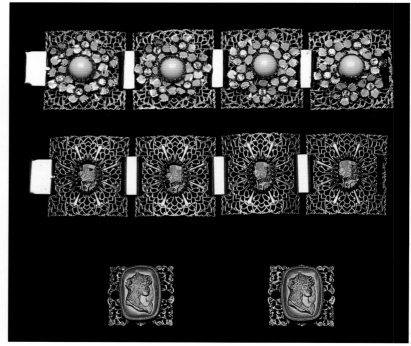

Metal filigree and plastic jewelry done three different ways. Goofy earrings have plastic center imprinted with an image of a Hawaiian girl. Bottom bracelet has Asian-themed center intaglios. (An intaglio is an incised design on glass or plastic.) Top bracelet has plastic center stones, surrounded by pink, painted, metal flowers. Unsigned. c.1950s. Earrings $10-15. Black bracelet $50-60. Pink bracelet $50-60.

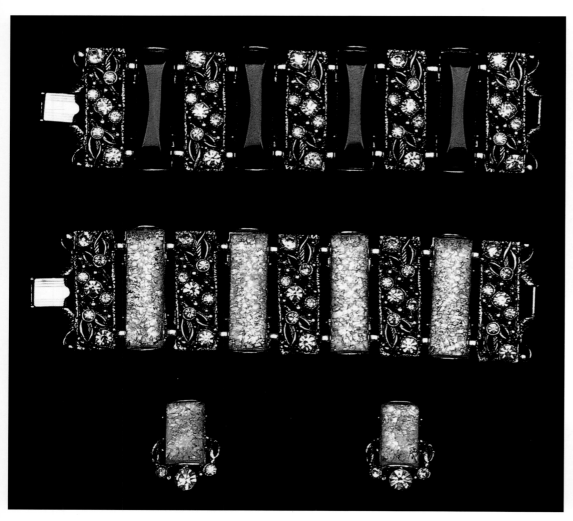

Large silvery confetti bracelet with matching earrings and black version of the same bracelet. c.1950s. Unsigned. Jewelry courtesy of Cheryl Killmer. Earrings $25-30. Each bracelet $65-85.

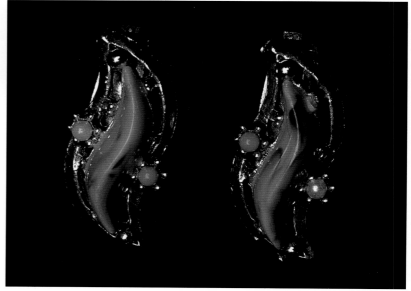

Turquoise swoop earrings. c.1950s. Unsigned. Jewelry courtesy of Cheryl Killmer. Earrings $10-15.

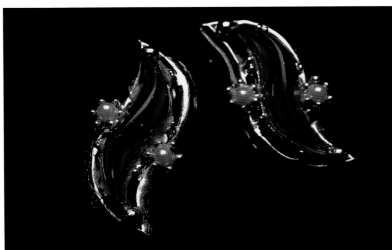

Tortoise shell colored version of the swoop earrings. c.1950s. Unsigned. $10-15

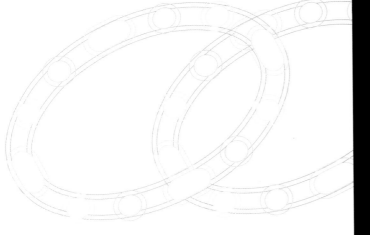

Three variations of this dainty set. From top, necklace bracelet and earrings in hot pink, orange, and light green. Lavender and white bracelet has same tiny plastic and painted metal leaves. Bottom bracelet, in shades of hot pink, orange, and light green, is quite similar, but has tiny plastic fruit shapes instead of leaves. c.1960s. Unsigned. Necklace $35-40. Each bracelet $30-35. Earrings $15-20.

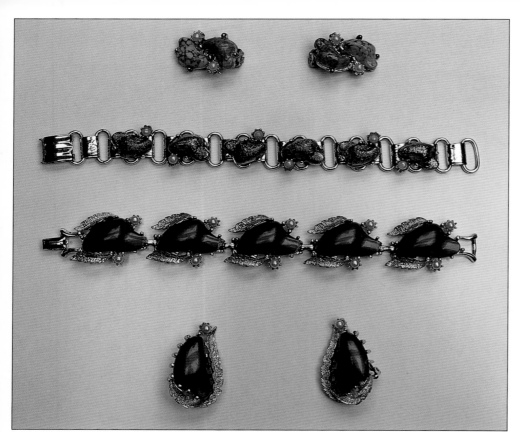

Amoebas! And lots of them! The shape is basically the same, but varies in size, texture, and color. Earrings with a mottled coral and turquoise look, unsigned. Bracelet of coated, iridescent amoebas, unsigned. Giant, brown amoebas used in earrings and bracelet by Coro. All c. 1950s. Unsigned earrings $20-25. Iridescent bracelet $35-40. Coro earrings $20-25. Coro bracelet $40-50.

More amoebas! Black amoeba bracelet with elaborate silver tone findings, unsigned. Black and red amoeba bracelet with a simple, but stunning setting, unsigned. Tortoise shell colored amoeba bracelet in a gold tone setting, unsigned. Coro earrings with double amoebas, large and small. All c.1950s. Unsigned, black bracelet $40-50. Unsigned red and black bracelet $50-60. Unsigned, tortoise shell colored bracelet $35-40. Black, Coro earrings $20-25.

Main Floor Savings

Specials as handy as they are handsome! Shop Tuesday 9 to 8:45

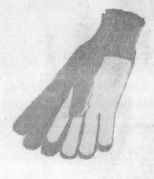

save *beautiful Exmoor shrugs*

Dayton's special purchase of cuddle-soft shrugs; excellent selection, light or dark colors. Knits are wool or Orlon; boucles, cotton-rayon blends. S, M, L. Sorry, no mail or phone orders filled. Reg. 3.98, **1.99**

DAYTON'S DRESS ACCESSORIES—MAIN FLOOR

save *exquisite stone jewelry*

Simulated stones: topaz, emerald, amethyst, bronze, jet, others; settings of antique gold, silver finish. Individual earrings; matched bracelet-earring sets, individually priced.

Earrings, **79c**   Bracelets, **1.69**
10% tax extra

DAYTON'S JEWELRY—MAIN FLOOR

save *smart on-campus gloves*

Just in the nick-of-time for school . . . sale of tuck-stitch wool knit gloves with pigskin palms. Sand, brown, navy, red, dark green, chamois, black, oatmeal. Smart on campus or just about anywhere. S, M, L. **1.95**

DAYTON'S GLOVES—MAIN FLOOR

save *"writing-home" notes*

Whether it's for away-at-school notes or any of your note-needs, you'll find this an excellent buy. Attractive Hallmark notes, four lovely floral or leaf designs. 12 to 18 to a box. Reg. 1.00, box **59c**

DAYTON'S STATIONERY—MAIN FLOOR

save *kitchen-fresh candy*

Dayton's delicious Maud Muller chocolates; 1 lb. box, reg. 1.15, *98c*; 2 lbs., reg. 2.20, *1.89* Cello-wrapped hard candies, 6 tempting flavors; 2 lbs., reg. 1.59, *1.35*. Old-fashioned Peanut Crunch, 1 lb., reg. 89c, *75c*

DAYTON'S CANDY SHOP—MAIN FLOOR

save *beautiful fall nylons*

Munsingwear Prima Donna . . . blush beige, or Taffy Apple, amber beige; daytime sheer, high twist for longer wear. 15 denier. Short (Iris) 8½-10; medium (Venus) 8½-11; long (Diana) 9½-11. Reg. 1.35, **1.07**

DAYTON'S HOSIERY—MAIN FLOOR

Dayton's department store (Minneapolis. MN) circular from Sept. 5, 1955, advertises "Simulated stones". Similar amoeba bracelet to pictured black and tortoise shell colored with elaborate findings examples. Earrings 79 cents and bracelet $1.79! Let's find a time machine and go back and buy some! Ad courtesy of Marshall Field's.

Flecks or Confetti

Ruby red, glittery bracelet. Unsigned. c. 1960s. $35-40.

Three varieties of large confetti/fleck bracelets with elongated rectangle shapes that are embedded with pearls and glitter. Unsigned. c. 1950s. Jewelry courtesy of Judy Levin Each $50-60.

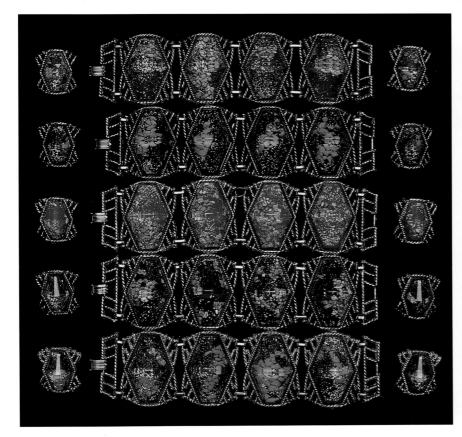

Hexagonally-shaped pieces filled with shells and glitter make these five bracelet and earrings sets quite dazzling. Unsigned. c. 1950s. Jewelry courtesy of Cheryl Killmer. Each Earrings Set $20-25. Each Bracelet $50-60.

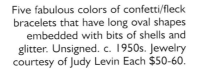

Five fabulous colors of confetti/fleck bracelets that have long oval shapes embedded with bits of shells and glitter. Unsigned. c. 1950s. Jewelry courtesy of Judy Levin Each $50-60.

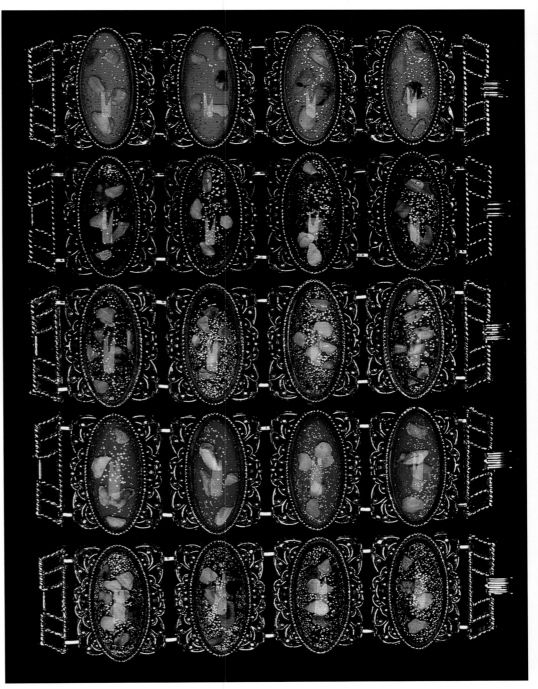

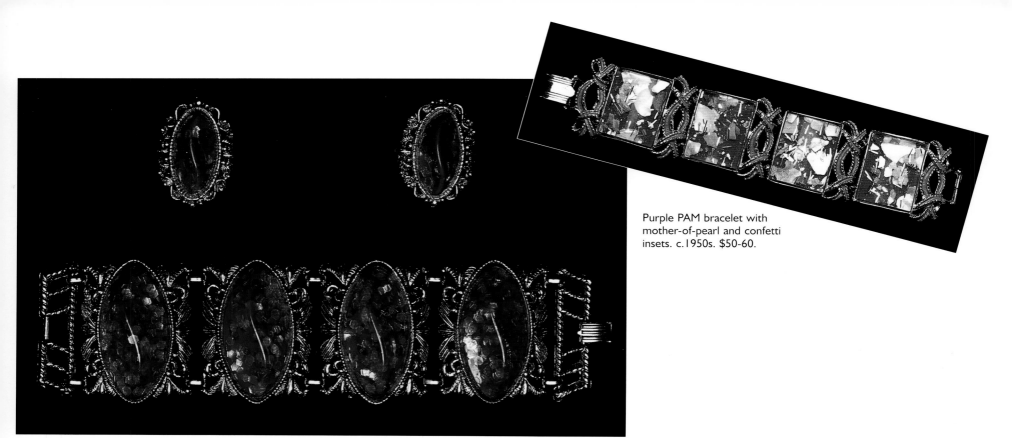

Purple PAM bracelet with mother-of-pearl and confetti insets. c.1950s. $50-60.

Spectacular, wave-shaped, red confetti bracelet and earrings. Unsigned. c. 1950s. Jewelry courtesy of Cheryl Killmer. Earrings $20-25. Bracelet $50-60.

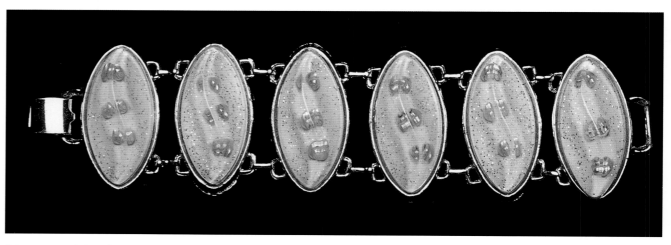

Echoing a sandy beach, this wave-shaped bracelet features three tiny iridescent shells atop silver glitter. Unsigned. c. 1950s. $50-60.

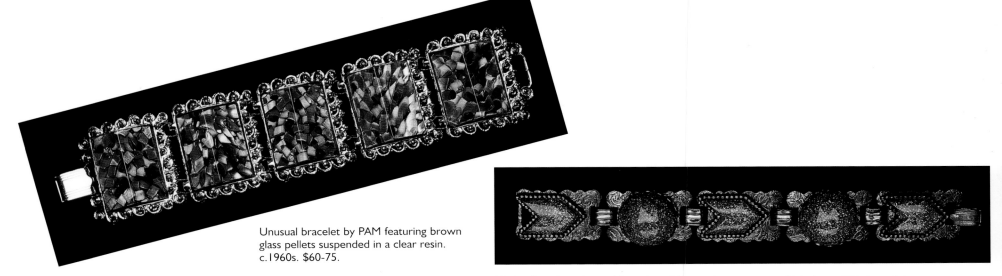

Unusual bracelet by PAM featuring brown glass pellets suspended in a clear resin. c.1960s. $60-75.

Native American themed bracelet with round, coppery, confetti pieces. This bracelet was probably a souvenir from the Western part of the United States. Unsigned. c. 1950s. $30-35.

Golden fleck bracelet with embedded bits of mother-of-pearl. Unsigned. c. 1950s. $30-35.

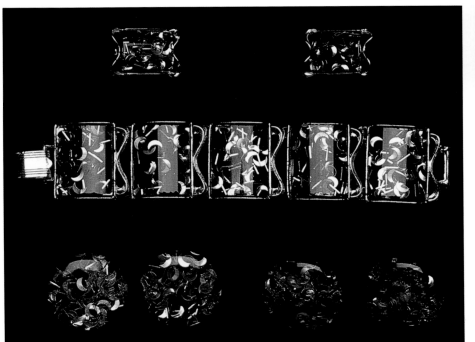

Glamorous black and gold. Top and center, bracelet with matching earrings, marked PAM. Bottom, nearly identical round earrings. Earrings on left unsigned and earrings on right signed Hobé. c. 1950s. Jewelry courtesy of Judy Levin and Carol Sullivan. Earrings on top $20-25. Bracelet $50-60. Unsigned round earrings $20-25. Round Hobé earrings $30-35.

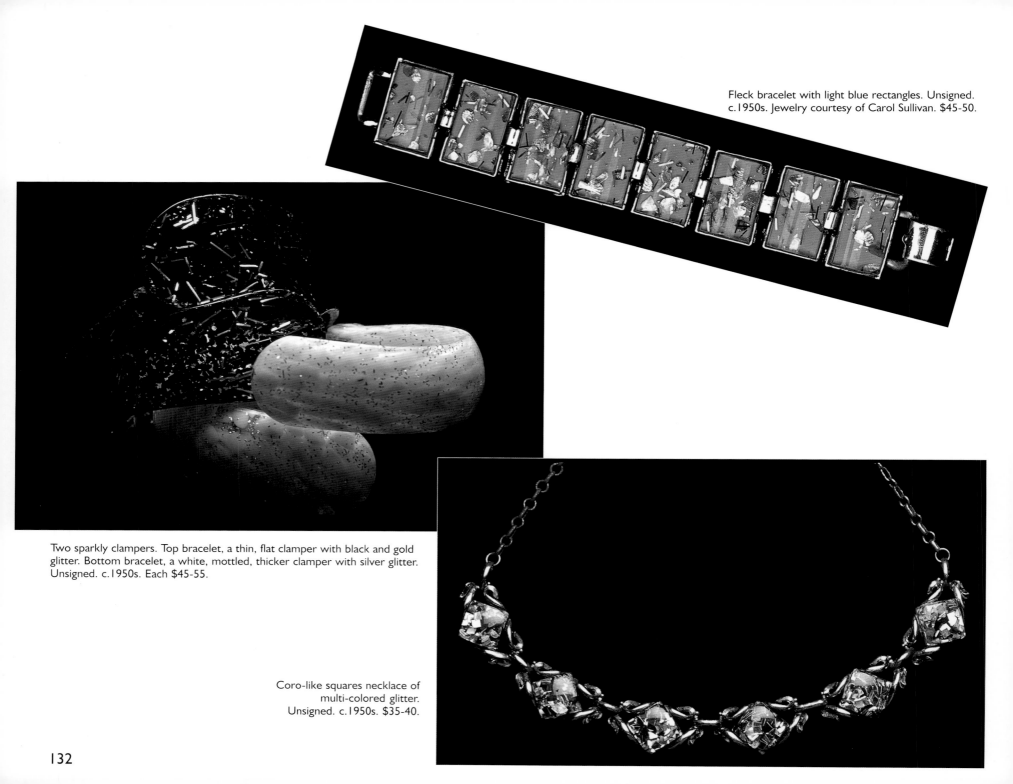

Fleck bracelet with light blue rectangles. Unsigned. c.1950s. Jewelry courtesy of Carol Sullivan. $45-50.

Two sparkly clampers. Top bracelet, a thin, flat clamper with black and gold glitter. Bottom bracelet, a white, mottled, thicker clamper with silver glitter. Unsigned. c.1950s. Each $45-55.

Coro-like squares necklace of multi-colored glitter. Unsigned. c.1950s. $35-40.

Extraordinary orange confetti cuff bracelet with matching earrings. Unsigned. c. 1950s. Jewelry courtesy of Carol Sullivan. Set $60-70.

What's black and white and red all over? No, not the newspaper, but this spectacular confetti clamper with matching earrings. Unsigned. c.1950s. Jewelry courtesy of Carol Sullivan. Set $70-75.

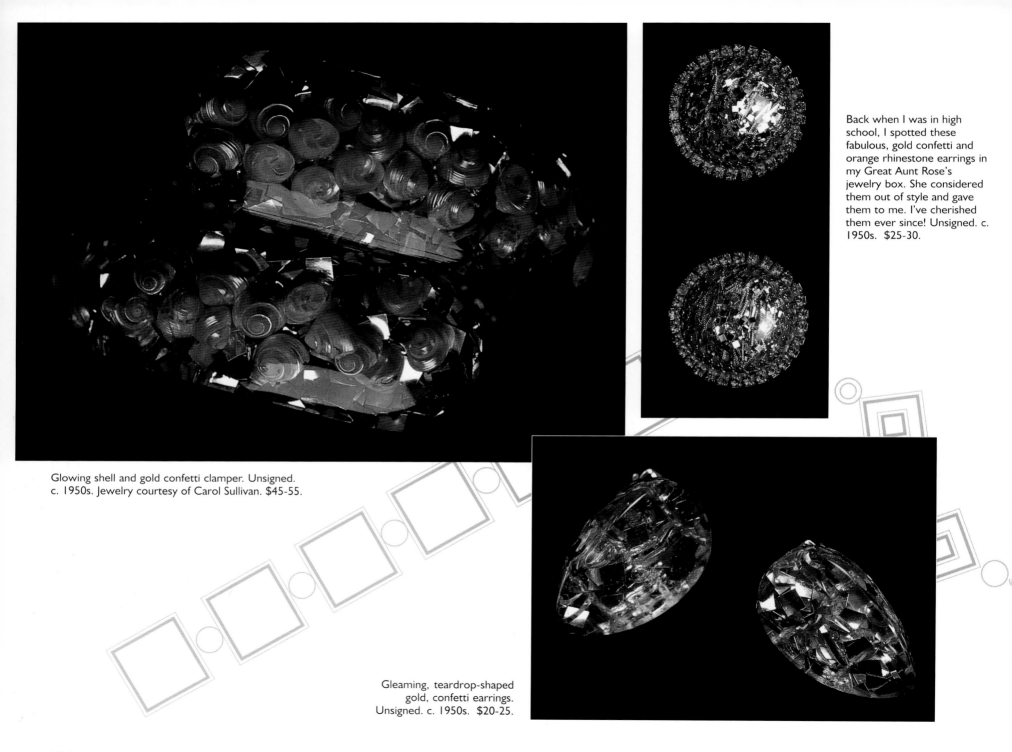

Back when I was in high school, I spotted these fabulous, gold confetti and orange rhinestone earrings in my Great Aunt Rose's jewelry box. She considered them out of style and gave them to me. I've cherished them ever since! Unsigned. c. 1950s. $25-30.

Glowing shell and gold confetti clamper. Unsigned. c. 1950s. Jewelry courtesy of Carol Sullivan. $45-55.

Gleaming, teardrop-shaped gold, confetti earrings. Unsigned. c. 1950s. $20-25.

Pretty in these pink and gold confetti earrings! Unsigned. c. 1950s. $20-25.

Two sets of button style confetti earrings in pink and blue. According to Jack Feibelman, former Director of Product Development for Coro, this type of confetti piece, with a solid colored back and a clear top of embedded glittery material, was most likely manufactured by Sobel Bros. of Perth Amboy, New Jersey. The plastic is a poured resin that embeds the mother-of-pearl chips and glitter. Unsigned. c.1950s. Each $25-30.

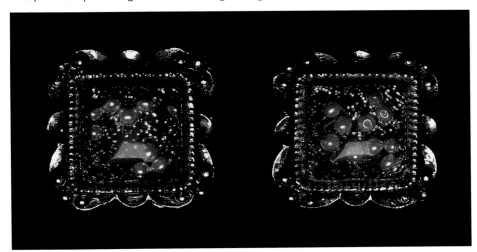

Seed pearls and gold glitter glimmer in these blue earrings. Unsigned. c. 1950s. $20-25.

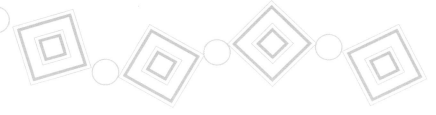

Fleck earring sets in purple and black. The purple set was originally a royal blue that faded out to purple. I discovered this when the plastic stone fell out of the setting and the back of the piece revealed its original color! Unsigned. c.1950s. Each $20-25.

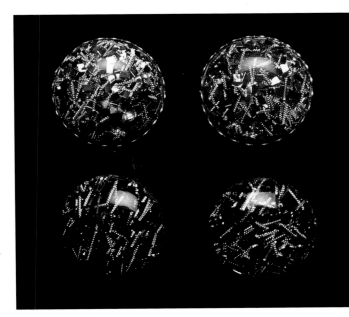

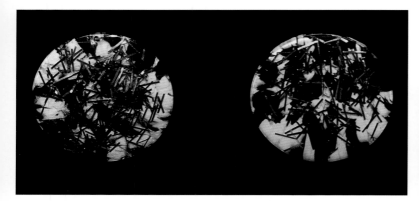

These earrings began life as buttons. I purchased them at an art fair, as such. It is interesting to note that the button manufacturers were using the fleck/confetti plastic, too. c.1950s. $25-30.

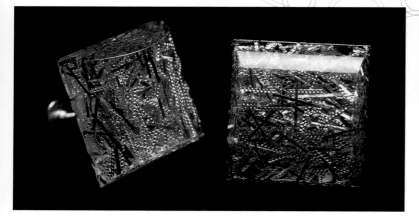

Screw-back earrings that look like they're embedded with silver sprockets. Unsigned. c.1950s. $15-20.

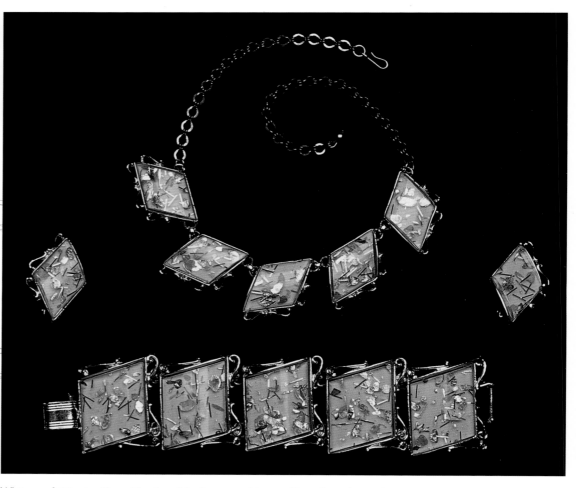

White confetti set with multi-colored flecks, comprising necklace, bracelet, and earrings. Unsigned. c.1950s. Jewelry courtesy of Carol Sullivan. Set $75-95.

Right:
Uncommon, dark green confetti necklace and earrings. Unsigned. c.1950s. Jewelry courtesy of Carol Sullivan and Judy Levin. Set $85-95.

136

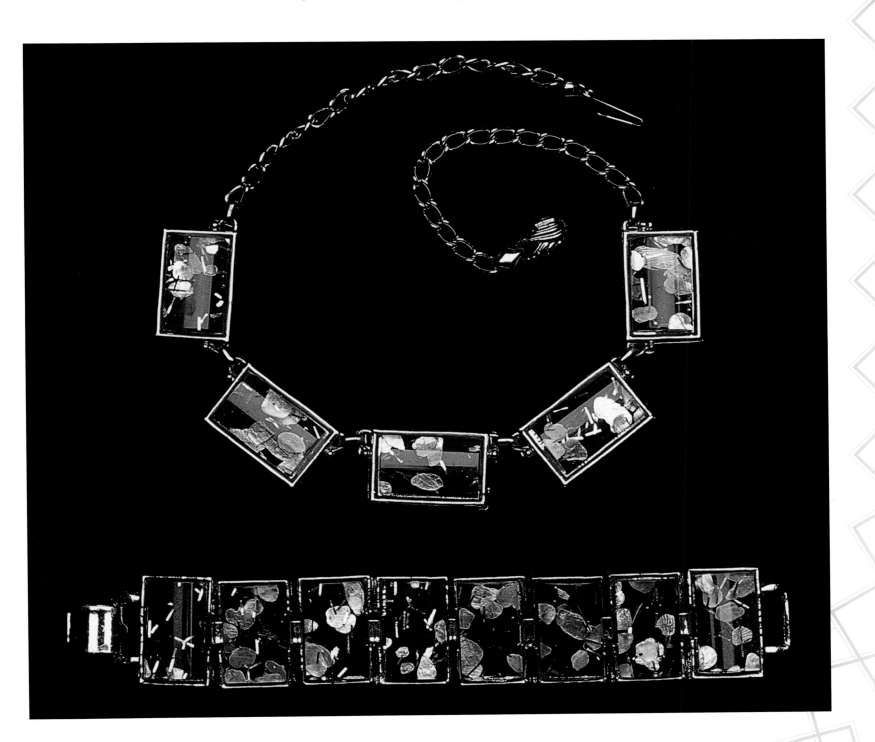

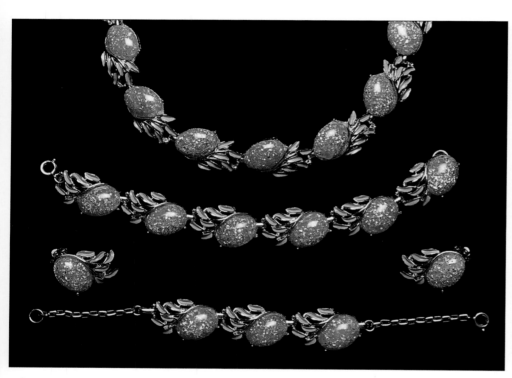

Apple green glittery, fleck set comprising necklace, two bracelets, and earrings. Bottom bracelet looks like it was newly composed out of old parts. Unsigned. c. 1950s. Set $85-95.

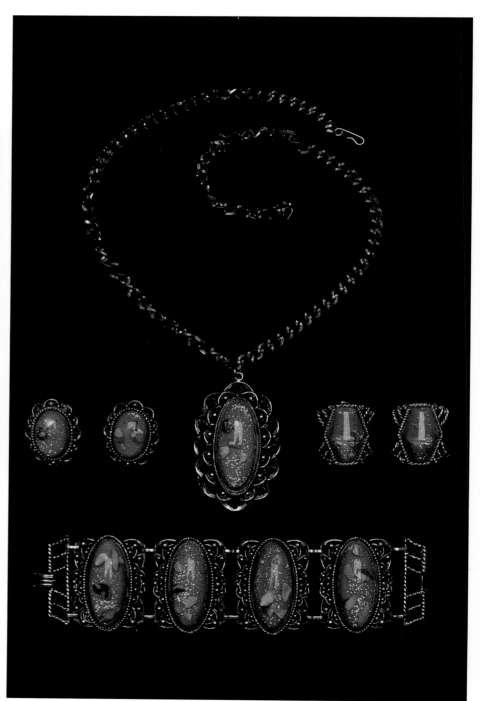

Grand, orange confetti set with pendant necklace, bracelet, and two styles of earrings. c.1950s. Unsigned. Jewelry courtesy of Judith Levin. Set $95-110.

4.
Styles That Sparkle

Bracelets

Chic, emerald green clamper bracelet with a pearly finish. Unsigned.
c.1950s. Jewelry courtesy of Cheryl Killmer. Set $45-55.

Bejeweled clampers. The Albert Weiss & Company originated this style of bracelet, which they called a "Bon Bon". According to Ed Sternberg, former Weiss salesman, the plastic in these bracelets was made by an umbrella handle manufacturer! If you look at the bracelet sideways, you can clearly see the bottom of the umbrella handle shape. The rhinestones used were always Swarovski. Sternberg says the bracelets were produced from about 1950 to 1960 and in 48 different colors and color combinations.

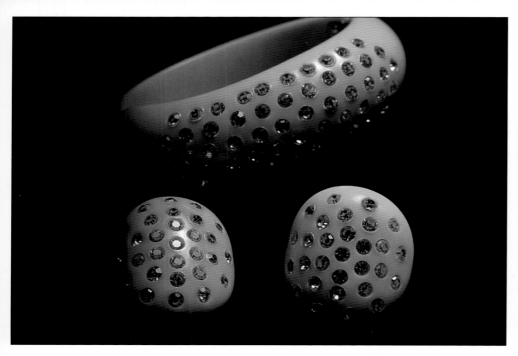

Lime green clamper with matching earrings, unsigned.
c.1950s. Jewelry courtesy of Cheryl Killmer. Set $150-175.

A triple threat of black clampers with two sets of matching earrings, unsigned. c.1950s. Jewelry courtesy of Cheryl Killmer. Each Bracelet $125-145. Earrings $35-45.

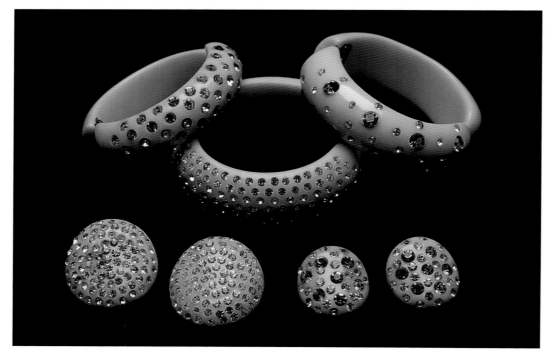

Trio of white clampers and a duo of earrings, unsigned. c.1950s. Jewelry courtesy of Cheryl Killmer. Each Bracelet $100-125. Earrings $35-40.

Luminous, emerald green rhinestones in a black setting. Unsigned. c. 1950s. Jewelry courtesy of Cheryl Killmer. Bracelet and earrings set $195-225.

A variety of shapes and textures come together in this elaborate, brown and taupe colored charm bracelet. Unsigned. c.1950s. $45-55.

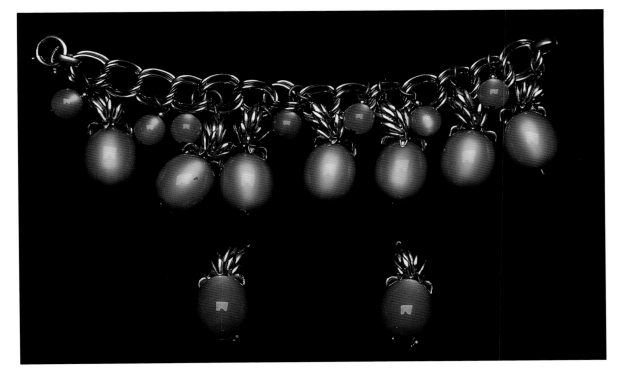

Satiny pineapples make a charming charm bracelet and earrings set by Napier. c.1960s. Jewelry courtesy of Cheryl Killmer. Set $125-135.

What do you suppose this designer was thinking when he or she added little gold tone pretzels to the mix on this charm bracelet! An unusual combination with a variety of beige dangles. Weird, but fun none-the-less! Unsigned. c.1950s. $40-50.

Vibrant green "pearls" adorn gold tone shells on this Coro charm bracelet. c.1950s. $40-50.

Delightfully, deco charm bracelet in green and orange. Unsigned. c.1950s. $35-40.

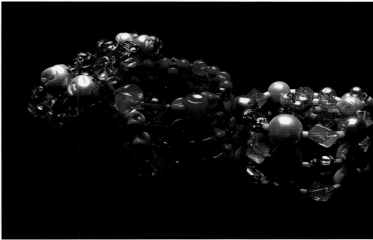

Manipulated (a strung piece, as opposed to being cast) bracelets of glass and plastic beads strung on curved wires. Unsigned. c.1950s. Purple bracelet $25-30. Beige and red bracelets $35-40.

Orange, yellow, cream, and white expandable bracelets. c. 1960s. With the exception of the yellow bracelet, all are marked simply "Hong Kong". Each $25-30.

Simple bracelets of manipulated plastic beads strung on curved wires. The green bracelet is probably a Richelieu Satinore. In the late 1940s, early 1950s, this bracelet sold for $3. Today, $35-40. The red and black bracelets are unsigned, c. 1950s. $25-30.

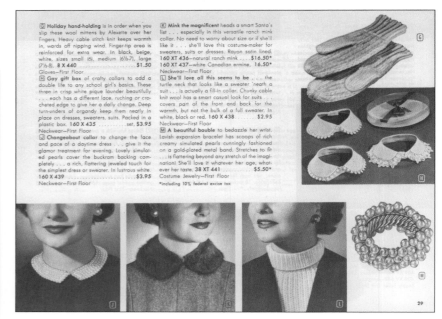

Note item "M" "A beautiful bauble to bedazzle her wrist". Marshall Field's department store (Chicago, IL) advertising circular Christmas, 1954. The expansion bracelet's $5.50 price tag included sales tax! Ad courtesy of Marshall Field's.

Watchband-style, expandable beaded bracelets in rust tones and black and white, marked "Japan." c.1960s. Each $35-40.

Six plastic beaded expansion bracelets (From top left) in white, black, red, blue, golden pearl, and silvery pearl. All are marked "Made in Japan", except for the black bracelet, which is marked "Kafin, New York" c.1950s. Each $35-45.

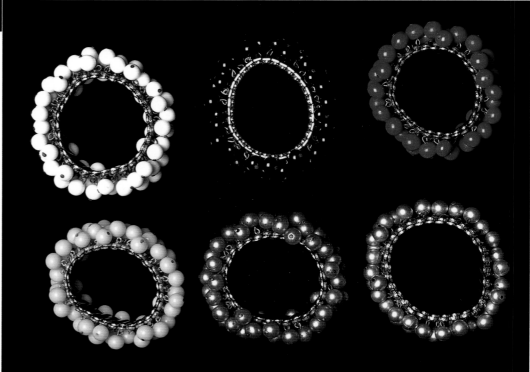

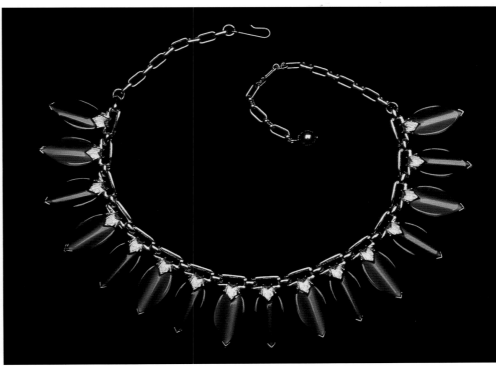

Feathery-style blue-green necklace. Signed Pakula. c.1950s.
Jewelry courtesy of Judy Levin. $35-45.

Large, bib style Goldette necklace
with faux coral accents. c.1960s.
Jewelry courtesy of Cheryl Killmer.
$65-75.

Marbled aqua-blue, texturized stones with pearls and AB rhinestones are featured in this necklace. (See similar bracelet on page 90) Unsigned. c.1950s. $30-35.

Green, leafy necklace by Leru (See similar bracelet and earrings on page 114).
c.1950s. Jewelry courtesy of Cheryl Killmer. $40-45.

Flowering hearts bloom on this necklace. Unsigned. c.1950s. $30-35.

This white Trifari necklace is missing two small plastic spheres. Without a duplicate piece to scavenge for parts, the spheres may be hard to match; however, if you choose to wear the item, make sure other plastic parts are secure and re-glue loose pieces. Consider removing a piece from the back of the necklace to reposition into the empty spot on the front. No one will know your secret! c. 1950s. $20-30 ($35-40 if in perfect condition).

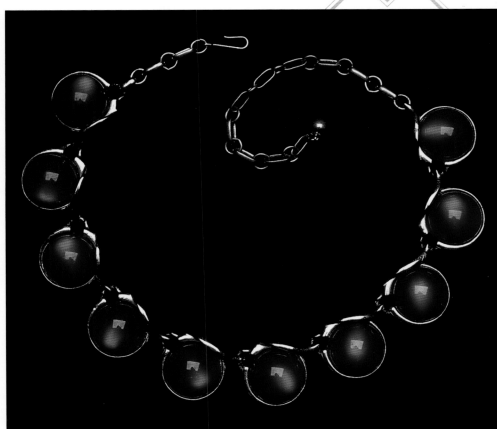

This necklace reminds me of ripe cherries! Signed Coro. c.1950s. courtesy of Judy Levin. $35-40.

Spherical, luminescent, grey plastic gems adorn this necklace. Unsigned. c.1950s. $30-35.

Earrings

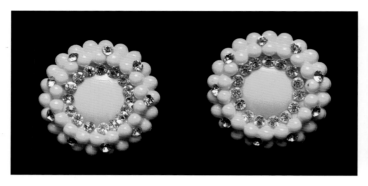

Sparkling, Swarovski rhinestones are readily apparent in these blue Weiss earrings. Undoubtedly, a matching earring for a Bon Bon bracelet. c. 1950-1960. $35-40.

Yellow and rhinestone sunflower earrings. Unsigned. c.1950s. $30-35.

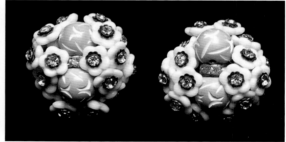

Little white flower bundles, each topped with a rhinestone and two carved, blue beads are featured in these Jonné earrings. c.1950s. $20-25.

Little orange bouquets of flowers for your ears. Marked "Hong Kong." c.1960s. $15-20.

Spectacular, two-tone green jelly earrings. Unsigned. c.1960s. $35-40.

Earrings consisting of tiny, blue plastic leaves. Unsigned. c.1950s. $15-20.

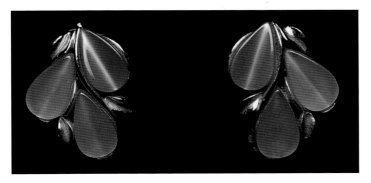

These lavender drops make attractive ear clips.
Unsigned. c.1950s. $10-15.

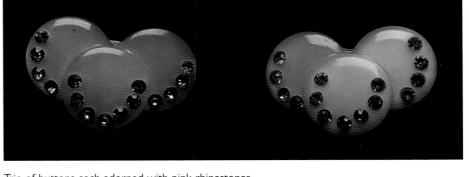

Trio of buttons each adorned with pink rhinestones
on these earbobs. Unsigned. c.1950s. $15-20.

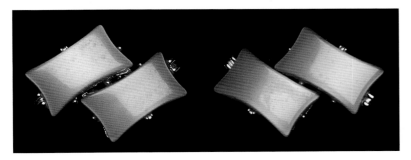

Mid-century style pink earrings. Unsigned. c.1950s. $10-15.

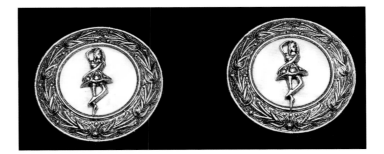

Ballerina earrings marked "Germany". Their thin, off-white, pearly
plastic material is probably cellulose acetate. c.1950s. $20-25.

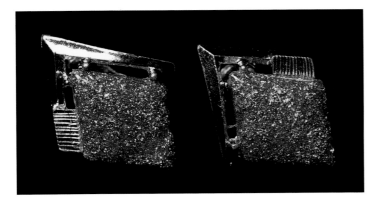

Diamond-shaped, copper colored fleck
earrings. Unsigned. c.1950s. $15-20.

Large, grey leafy earrings by Lisner. c.1960s. $20-25.

1957 Trifari ad, "She can't see a thing but jewels by Trifari".
Notice the earrings featured in item "D" on this ad.

Newspaper ad for Daytons's department store (Minneapolis, MN), June 1955, featuring "Ear Whispers by Coro". Note the illustration of the round earrings with the floral detail in the middle of the ad. Ad courtesy of Marshall Field's.

Unsigned, big, round, pearly earrings look very similar to those featured in the Coro ad. $20-30.
Copying was a fairly common occurrence. A 1955 federal lawsuit by Trifari had landmark repercussions, copyrighting fashion jewelry design as a work of art.

Unsigned knock-off of the earrings featured in item "D" of the Trifari ad. Notice the incredible similarity to the Trifari earrings. c.1950s. $20-25.

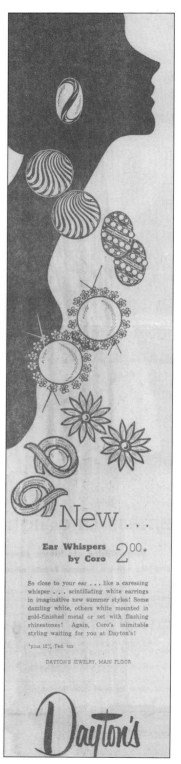

New...

Ear Whispers $2.00*
by Coro

So close to your ear . . . like a caressing whisper . . . scintillating white earrings in imaginative new summer styles! Some dazzling white, others white mounted in gold-finished metal or set with flashing rhinestones! Again, Coro's inimitable styling waiting for you at Dayton's!

*plus 10% Fed. tax

DAYTON'S JEWELRY, MAIN FLOOR

Dayton's

Golden leaf pin with a single rhinestone. Unsigned. c. late1940s. $35-40.

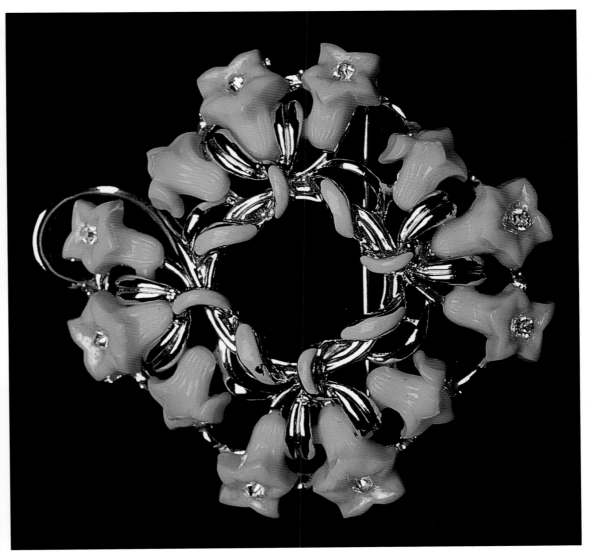

Yellow, daffodil pin by Star. c.1950s. $35-40.

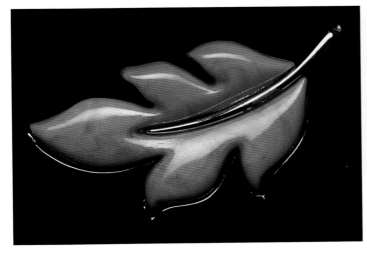

"Autumn Splendor" pin by Sarah Coventry. c.1970s. $25-30.

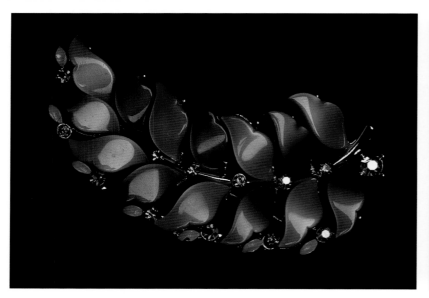

Like a feather! Beige with gold and orange rhinestones. Signed Lisner. c.1950s. courtesy of Judy Levin. $35-45.

Two elegant, pins comprised of grey, opalescent, plastic stones and clear rhinestones. Signed Coro. c.1950s. Jewelry courtesy of Cheryl Killmer. Smaller pin $65-75. Larger pin $100-125.

Flower power in grey. Unsigned. c. 1960s. $25-30.

Green grapes pin. The little glowing fruits are actually three different shades of green. Unsigned. c. 1960s. $35-40.

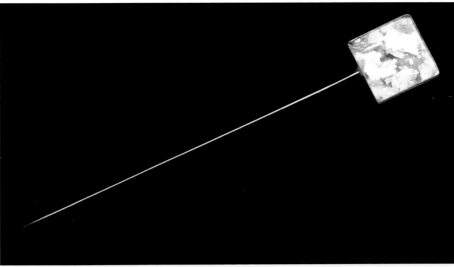

Fleck stick pin with embedded pieces of mother-of-pearl. Unsigned. c.1970s. $15-20.

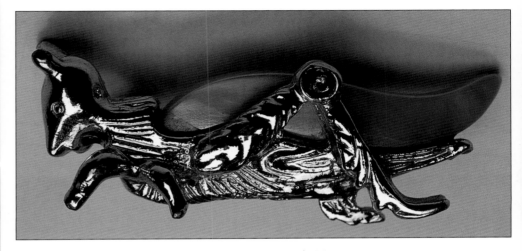

Tiny green grasshopper pin. Adorable. Unsigned. c. 1950s. $20-25.

Blue and white leaves pin. Unsigned. c.1950s. $25-30.

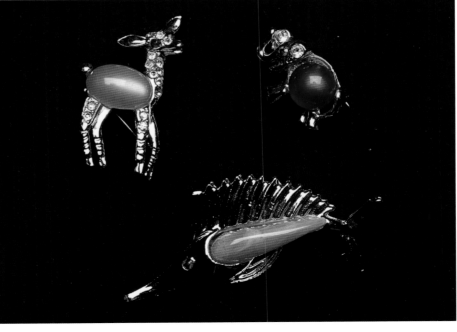

Little animal pins. From top left, deer with pink, plastic stone. Elephant with red, plastic stone. Swordfish with pink plastic stone. All unsigned. c.1950s. Each $15-20.

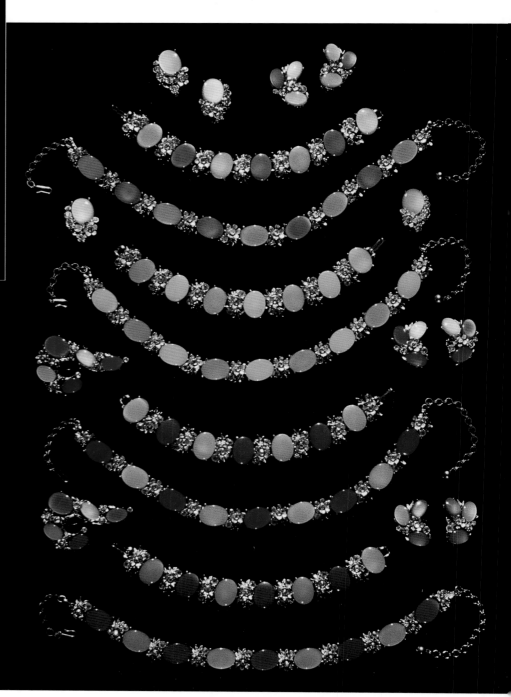

You'll go quackers for this painted, plastic duckies pin with clear rhinestones. Unsigned. c. 1940s. Jewelry courtesy of Judith Levin. $75-95.

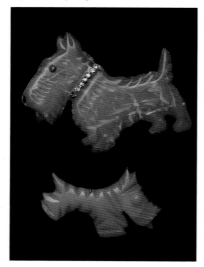

Woof! Two pink, plastic Scottie dog pins. Unsigned. c. 1940s. Jewelry courtesy of Judith Levin. Large pin $75-100. Small pin $50-65.

154

Green, leafy Lisner complimentary sets. Necklace and bracelet are one design, while pin and earrings are another. c. 1960s. Jewelry courtesy of Cheryl Killmer. Necklace $50-60. Bracelet $50-60. Pin $45-55. Earrings $25-30.

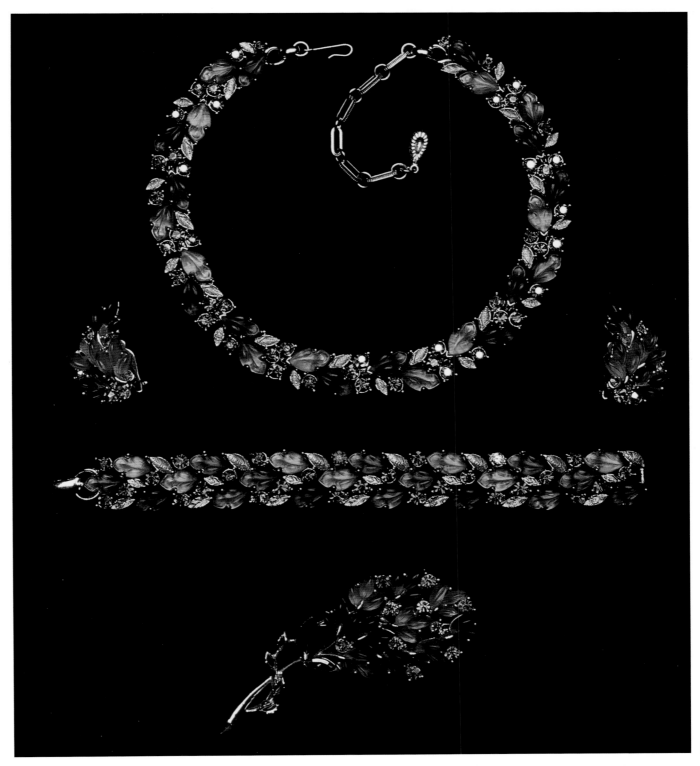

Left:
B.S.K. parures in four colorful varieties. c. 1950s. Jewelry and photo courtesy of The Francesca Medrano Collection. Each necklace $35-40. Each bracelet $30-35. Each set earrings $15-20. Each pin $30-35.

Green and grey "wishbones" bracelet with center rhinestones. Earrings are solid green. Unsigned. c.1950s. Set $35-40.

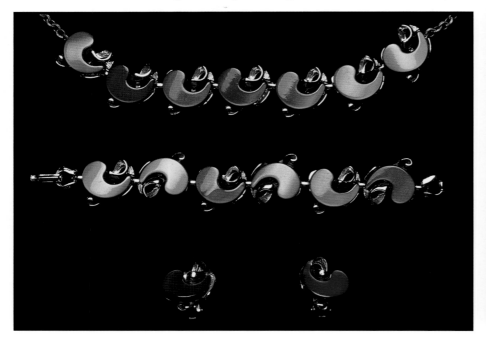

Green curlicues parure. Unsigned. c. 1950s. Set of necklace, bracelet, and earrings $50-60.

Green and blue parure. Findings mirror the plastic parts. Unsigned. c. 1950s. Set of necklace, bracelet, and earrings $50-60.

156

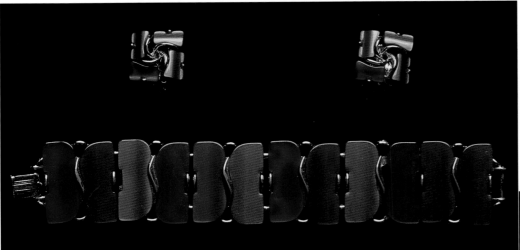

Orange, green, and gold bracelet with plastic parts that resemble the letter "B" and matching earrings. Unsigned. c. 1950s. Set $50-60.

Shiny, golden yellow cabochons ringed with floral findings. Unsigned. c.1950s. Jewelry courtesy of Cheryl Killmer. Parure of necklace, bracelet, and earrings $60-75.

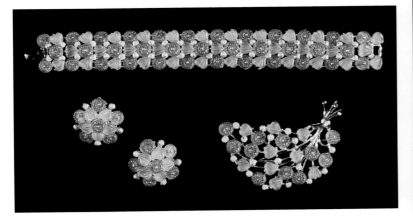

Jelly floral set in two shades of orange. Unsigned. c.1960s. Jewelry and photo courtesy of The Francesca Medrano Collection. Demi-parure of bracelet, earrings and pin $75-100.

157

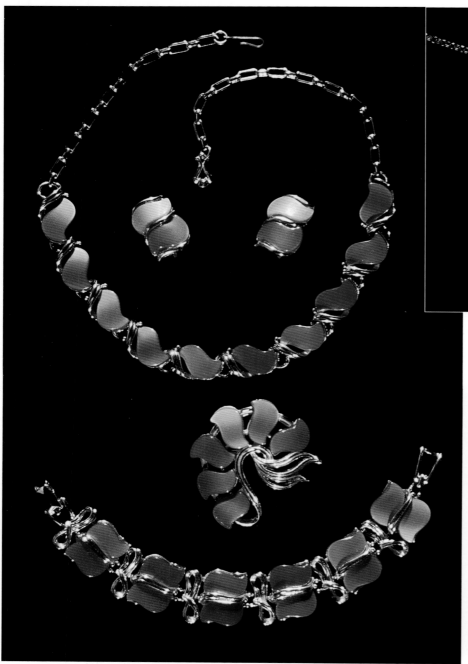

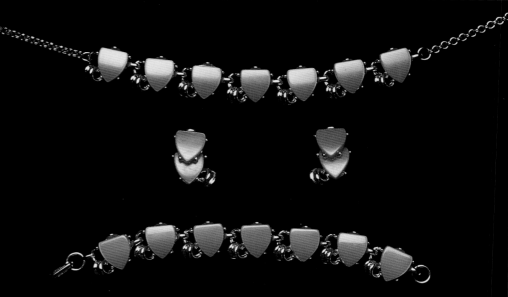

Orange shield shapes decorate this parure. Unsigned. c.1950s.
Set is composed of necklace, bracelet, and earrings $50-75.

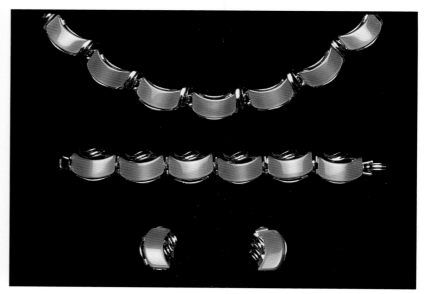

Full parure of necklace, bracelet, pin, and earrings in rust and gold by Star. c.1950s.
Jewelry and photo courtesy of The Francesca Medrano Collection. $75-100.

Butterscotch-colored necklace, bracelet, and earrings.
Unsigned. c.1950s. Set $50-60.

Large variety of Claudette pieces with tiny brown, orange, and yellow leaves. Set includes duo of necklaces, bracelet, pin, and two earring styles. c.1950s. Jewelry and photo courtesy of The Francesca Medrano Collection. Entire parure $95-120.

Graceful tan and brown parure. Unsigned. c.1950s. Jewelry and photo courtesy of The Francesca Medrano Collection. Set of necklace, bracelet, and earrings $60-80.

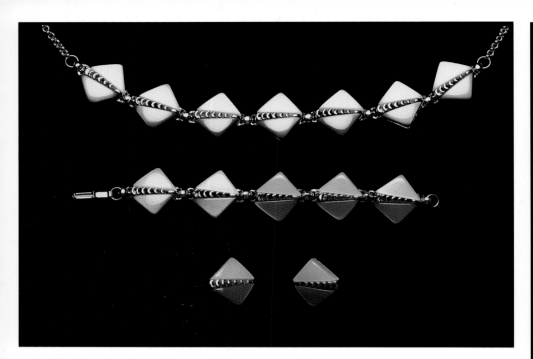

Necklace, bracelet, and earrings of two-tone triangles
in beige and brown. Unsigned. c.1950s. Parure $50-60.

Sleek grey parure by Charel. c.1950s. Jewelry and
photo courtesy of The Francesca Medrano Collection.
Set of necklace, bracelet, and earrings $75-100.

Olive and soft green hued necklace
and bracelet with little sharks-teeth
shaped parts. No, they don't bite.
Unsigned. c. 1950s. Jewelry courtesy
of Cheryl Killmer. Set 60-75.

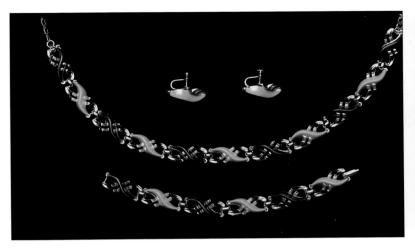

Swoosh-shaped set with necklace, bracelet, and screw-back earrings in aqua and brown. Unsigned. c.1950s. Entire Parure $50-75

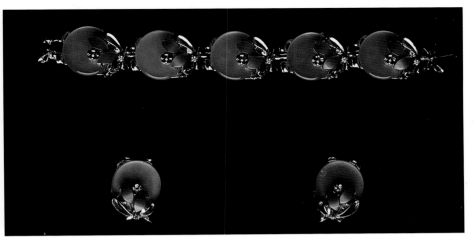

Girly bracelet and earrings set of blue circles, topped with a silver tone bow and flower with tiny contrasting blue leaves. Marked Andrea on bracelet only. Each link is marked very lightly on the back. Examine jewelry closely for marks that are light or are in peculiar places. Sometimes not all pieces are marked, so when you find the bracelet mate to a pair of unsigned earrings, you may finally discover the manufacturer. c.1950s. Demi-parure $50-60.

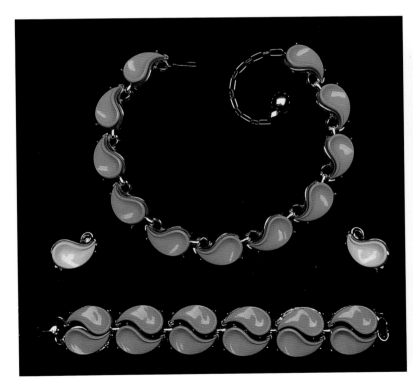

Exclamation, it's commas by Lisner! Necklace and bracelet in aqua, earrings in pink. c.1950s. Jewelry courtesy of Cheryl Killmer. Necklace $40-50. Bracelet $35-40. Earrings $25-30.

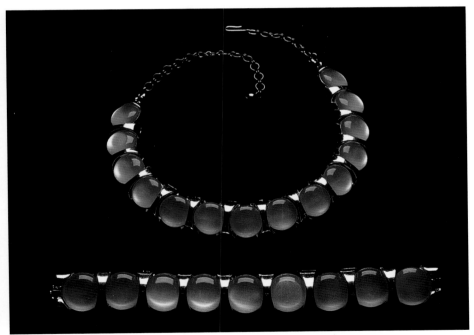

Florescent blue disks set into swank silver tone findings. Unsigned. c.1960s. Necklace and bracelet set $60-75.

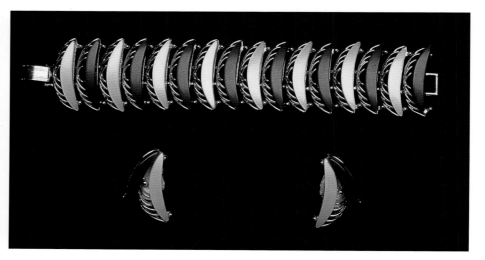

Smart navy blue and white bracelet and earrings. Unsigned. c.1950s. Set $50-60.

Lisner loved their leaves and so do I! Interchanging green and blue leaves on necklace and bracelet. c.1960s. Set $60-75.

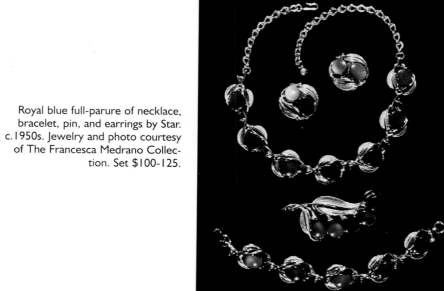

Royal blue full-parure of necklace, bracelet, pin, and earrings by Star. c.1950s. Jewelry and photo courtesy of The Francesca Medrano Collection. Set $100-125.

Necklace, bracelet, and fur clip of satiny wine-colored stones set into rhinestone-encrusted findings. Fur clip and bracelet signed Coro. Necklace unsigned. c. 1940s. Jewelry courtesy of Judy Levin. Set $150-200.

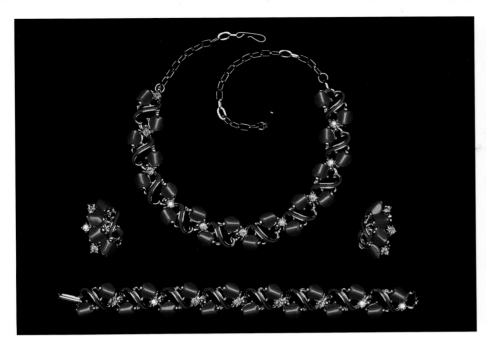

Festive plastic "red berries" enhance this necklace, bracelet, and earrings set. Unsigned. c. 1950s. Jewelry courtesy of Cheryl Killmer. Parure $60-75.

Necklace and earrings of red flame-shaped parts surrounded by tiny floral findings. Unsigned. c. 1950s. Jewelry courtesy of Cheryl Killmer. Set $65-75.

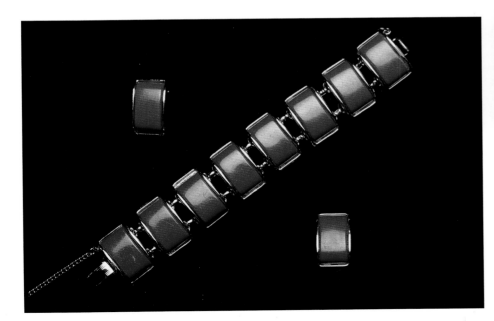

Curved red oblongs ornament this bracelet and earrings. Unsigned. c.1950s. Set $40-50.

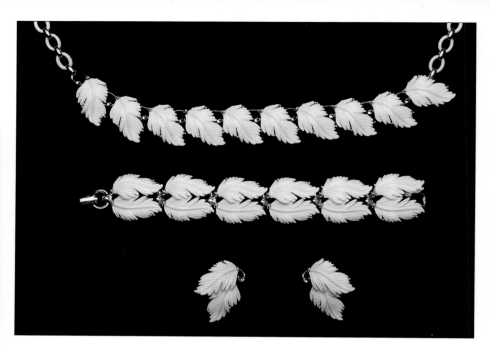

Pale pink, feathery Lisner leaves. c.1950s. Necklace, bracelet, and earrings $70-85.

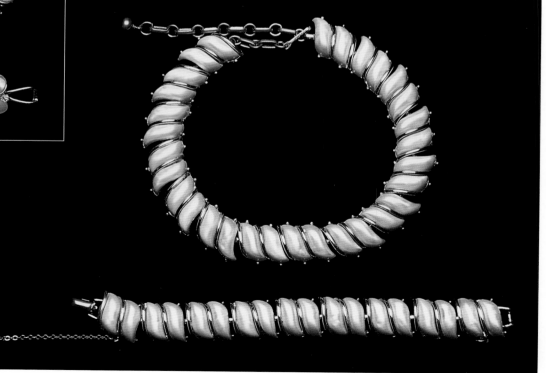

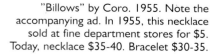

Who says you have to have green clovers to be lucky. This pink clover set with do the job just as well! Unsigned. c.1950s. Necklace, bracelet, and earrings $75-85.

"Billows" by Coro. 1955. Note the accompanying ad. In 1955, this necklace sold at fine department stores for $5. Today, necklace $35-40. Bracelet $30-35.

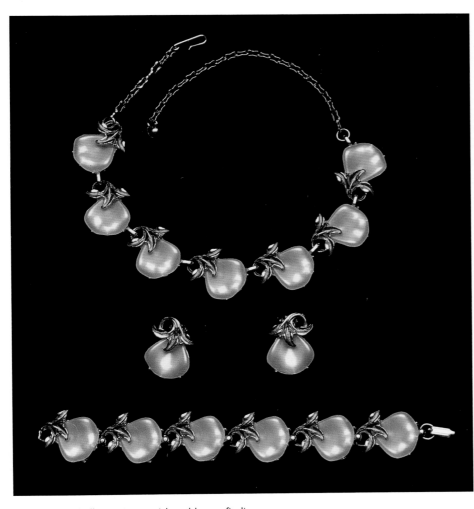

Pink, plastic shells, set into swirly gold tone findings.
Unsigned. c.1950s. Necklace, bracelet, and earrings $65-75.

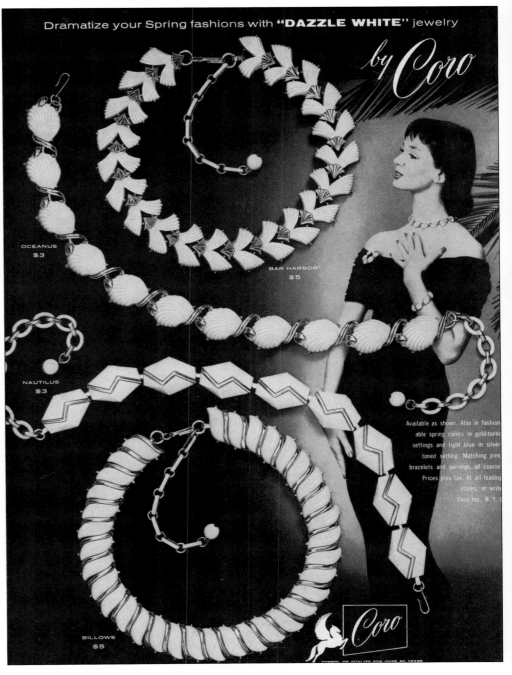

Dramatize your Spring Fashions with "Dazzle White" jewelry by Coro. 1955.
Ad courtesy of Pat Seal, Treasures from Yesterday, Fort Worth, Texas.

165

Novelty Jewelry

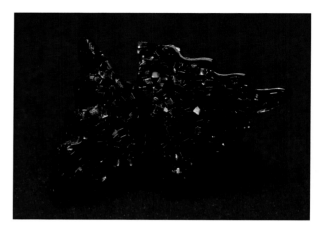

Merry Christmas! Confetti style! Gold and green confetti Christmas tree earrings are a seasonal delight. Unsigned. c.1950s. $25-35.

The author's parents and a family friend in Hawaii, c. early 1950s. Jewelry shown is touristy pink and pearl hula-girl charm bracelet with Hawaii medallion and green plastic south-seas look plastic bead necklace. Both unsigned. Both c.1950s. Green necklace courtesy of Cheryl Killmer. Green necklace $35-40. Charm bracelet $35-40.

Circa 1950s Florida tablecloth highlights whimsical "tourist" charm bracelet embellished with plastic beads and cabochons. Charms include an airplane, ocean liner, the Eiffel Tower, and the Arc de Triomphe. Unsigned. c.1960s. $75-95.

Each Christmas season during the late 1960s, my family and I stayed at a long-gone, mid-century gem of a hotel in Scottsdale, Arizona, "The Safari Inn". I coveted jewelry like this from the hotel gift shop. My mother always said I didn't need it, and I still don't need it, but I sure do want it! Souvenir necklace and bracelet with stamped, copper plated findings and pink confetti stones. Unsigned. c. 1960s. Set $30-40.

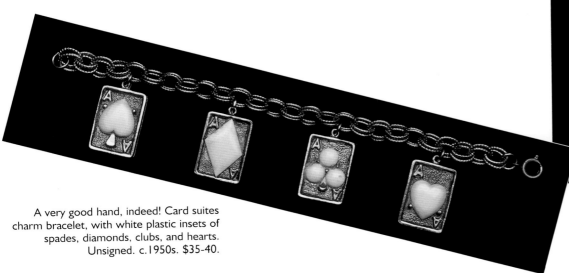

A very good hand, indeed! Card suites charm bracelet, with white plastic insets of spades, diamonds, clubs, and hearts. Unsigned. c.1950s. $35-40.

Rare Trifari sweater clip in turquoise. c. 1950s. Jewelry courtesy of Cheryl Killmer. $30-40.

The Forbidden Fruits

"The Forbidden Fruits" lush pineapple pin and earrings set. The rhinestones are sunk directly into the acrylic bodies of these magnificent specimens. We know they were marketed as "The Forbidden Fruits", but unfortunately we do not know their manufacturer. The jewelry was never signed, always paper tagged. c.1950s-1960s. Jewelry courtesy of Cheryl Killmer. Set $150-175.

Veggie delight! Eggplant pin and radish pin and earrings. The Forbidden Fruits. c. 1950s-1960s. Jewelry courtesy of Cheryl Killmer. Eggplant pin $125-145. Raddish pin and earrings set $150-175.

Black plastic fruit pin with white, faceted glass stones. The Forbidden Fruits. c. 1950s-1960s. Jewelry courtesy of Cheryl Killmer. $85-95.

Plastic lemon pin with triangular rhinestone detail. The Forbidden Fruits. c. 1950s-1960s. Jewelry courtesy of Cheryl Killmer. $85-95.

Ripe, yellow pear pin and earrings in "The Forbidden Fruits". c. 1950s-1960s. Jewelry courtesy of Cheryl Killmer. Set $135-145.

5.
Condition, Care & Repair

Lovely, Lisner bracelet was purchased for 90 cents! Although it's missing a plastic stone, it could be used for replacement parts for another piece. The price was definitely right! I, however, had a costume jewelry repair person remove the dried glue in the finding and then give it a fresh coat of gold leaf. I wear and enjoy it 'as is'!

Buying for Parts

It is not always necessary to purchase a piece of costume jewelry that is the perfect specimen. Broken clasps and common white or AB rhinestones can easily be replaced. Look for broken pieces with unusual rhinestones or plastic parts. Sometimes, purchasing a beyond repair piece for the right price will give you the perfect replacement stone for another imperfect treasure. Because this jewelry is of a vintage nature, the glue used to secure rhinestones and plastic parts can become dry and the pieces can easily fall out. Hence, if you have a piece which you enjoy wearing, consider buying a twin that is a less perfect example. You will be able to cannibalize the second piece for its parts, should a stone mishap occur.

Water & Hand Lotion: Jewelry's Worst Enemy

When cleaning a piece of jewelry, gently wipe with a soft, dry cotton cloth. A cotton swab or soft, dry toothbrush can be used for further detailing. Using water as a cleaning agent will loosen old glue and will deaden old rhinestones by blackening their foil backing. On non-plastic costume jewelry, a small amount of glass cleaner on a paper towel will also aid in reconditioning.[194] One must take particular care when handling old plastics, as common household products can act as solvents that can lead to their degradation. Keep hands clean when handling plastic jewelry, as even hand lotion can act as a solvent for plastics. The lotion penetrates the plastic material and will begin to dissolve it. A common example of this process is when, due to greasy foods, plastic food storage containers become sticky from use. Nail polish remover (acetone) will destroy Celluloid (cellulose nitrate), cellulose acetates, and acrylics. Perfume, vinegar, and ammonia based household cleaners can also attack plastics. The concentration and duration of these common substances will determine how harmful the effect upon the particular plastic. "Bloom", a white, chalky look to a plastic material, may be the first sign of degradation. A strong smell can also be indicative of a problem, particularly in Celluloid (cellulose nitrate) plastics. A vinegar smell is emitted when the camphor molecules evaporate, leading the cellulose to break down, thereby releasing nitrogen dioxide gas. The loss of the plasticizer, the substance that makes the plastic softer and more flexible, and the resulting chemical release can infect other Celluloid-type plastic articles.[195, 196] An infected Celluloid-type piece should be quarantined from other like pieces in the collection. Loss of plasticizer in a cellulose acetate piece will cause stiffening and distortion. Storage in bright light, airtight conditions, or extreme heat can also negatively affect plastics. If you desire to wrap your plastic jewelry, it should be done loosely, to promote ventilation, with uncolored, acid-free tissue paper. One must be cautious with the storage of decomposing Celluloid (cellulose nitrate) articles in cellulose-based storage materials (paper, cardboard), as the release of the nitrogen dioxide gas can cause these materials to become "nitrated," and therefore flammable.[197, 198]

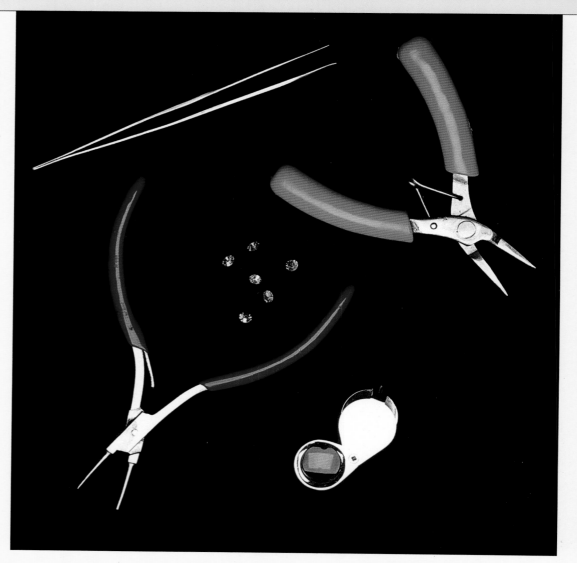

Helpful jewelry tools for the maintenance of your collection.

Basic Jewelry Tools

Handling small repairs, such as replacing a simple stone or adjusting a bent prong, are an essential part of collecting costume jewelry. Some of the basic jewelry tools that one will need to maintain a collection are: a magnifying glass or jeweler's loupe, a tweezers or needle-nosed pliers, and a tube of clear craft or jeweler's glue. Using a super glue or crazy glue-type product is not recommended. A small scissors, exact-o-knife or wire cutters, and safety goggles are more advanced tools that may be helpful to your care and repair needs. It is also helpful to have a stash of basic white or AB rhinestones and perhaps some seed pearls. All of these items can be purchased at your local hardware and/or craft store. You can use the loupe to see what you're doing or to, more easily, iden-tify a maker's mark. The pliers can be used to adjust a stray prong back into its proper place or to position a replaced rhinestone. Take great care when re-gluing old plastic parts to keep the glue solely on the back of the piece, as stray glue can cause damage to plastics. A simple toothpick can be helpful in aiding the precision of gluing.[199] More complicated repairs should be attempted by a jeweler, particularly one who is adept at fixing costume jewelry. There are many chapters of the Vintage Fashion & Costume Jewelry club (VFCJ) across the country. Joining a costume jewelry collecting club will allow you to share information regarding all aspects of this hobby.[200] Through the resources of the collector's club, one can further access sources for costume jewelry care and repair.

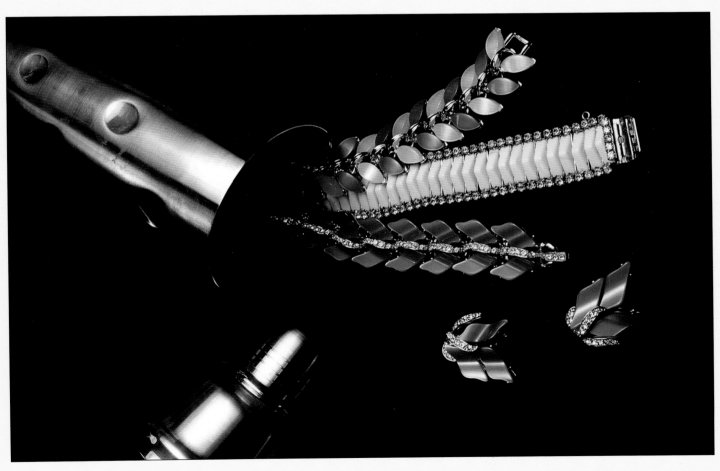

Cheers and happy hunting!

End Notes

1. Interview with Irving Wolf, former President of Trifari, 2003.
2. McCormick, Elsie, "Merchants of Glitter", *Saturday Evening Post*, May 31, 1947, pp. 40-41, 144-146.
3. Interview with Richard Norton II of Richard Norton Inc., 2003.
4. Weisberg, Alfred M., *Why Providence?* Providence, Rhode Island: Technic, Inc., 1992, p. 1.
5. Weisberg, Naida D. *Diamonds are Forever, but Rhinestones are for Everyone!* Providence, Rhode Island: The Providence Jewelry Museum, 1999, p. 12.
6. *Findings, The Newsletter of The Providence Jewelry Museum,* Volume 1, Issue 1, Autumn 2001, p. 3.
7. McCormick, Elsie, "Merchants of Glitter", *Saturday Evening Post*, May 31, 1947, pp. 40-41, 144-146.
8. "Costume Jewelry", *Fortune Magazine*, December 1946, pp. 140-145, 214, 217.
9. "Science Now: The Smell of a Hard Shower", *The American Association for the Advancement of Science Website,* URL: http://www.sciencenow.sciencemag.org/archives.101997.shtml, Friday, October 17, 1997.
10. DuBois, J. Harry. *Plastics History U.S.A.* Boston, Massachusetts: Cahners Books, 1972, pp. 58-59.
11. Lauer, Keith and Robinson, Julie. *Celluloid: Collector's Reference and Value Guide.* Paducah, Kentucky: Collector's Books, 2001, p. 15.
12. DuBois, J. Harry. *Plastics History U.S.A.* Boston, Massachusetts: Cahners Books, 1972, pp.38-57, 266.
13. Lauer, Keith and Robinson, Julie. *Celluloid: Collector's Reference and Value Guide.* Paducah, Kentucky: Collector's Books, 2001, pp. 8-14, 27.
14. Ibid., p. 62.
15. "Time 100: Chemist Leo Baekeland", *Time Magazine Online,* URL: http://www.time.com/time/time100/scientist/profile/baekeland.html, 2003.
16. Battle, Dee and Lesser, Alayne. *The Best of Bakelite and Other Plastic Jewelry.* Atglen, Pennsylvania: Schiffer Publishing, Ltd., 1996, p. 7.
17. DuBois, J. Harry. *Plastics History U.S.A.* Boston, Massachusetts: Cahners Books, 1972, p. 104.
18. Robins, Natalie and Aronson, Steven M.L., *Savage Grace.* New York, New York: Dell Publishing Co., Inc., 1985, p. 118.
19. Interviews with Maureen Reitman, Senior Managing Engineer, of Exponent Inc., 2003, 2004.
20. DuBois, J. Harry. *Plastics History U.S.A.* Boston, Massachusetts: Cahners Books, 1972, pp. 4, 33-34.
21. Baker, Lillian. *Plastic Jewelry of the Twentieth Century.* Paducah, Kentucky: Collector's Books, 2003, p. 190.
22. DiNoto, Andrea. *Art Plastic: Designed for Living.* New York, New York: Abbeville Press, 1984, p. 20.
23. "History: Celanese, a Company with Tradition" Website, URL: http://www.celanese.com/index/about_index/history-1999-1980/history-1918-1863.htm, 2004.
24. DuBois, J. Harry. *Plastics History U.S.A.* Boston, Massachusetts: Cahners Books, 1972, pp. 268-270.
25. Lauer, Keith and Robinson, Julie. *Celluloid: Collector's Reference and Value Guide.* Paducah, Kentucky: Collector's Books, 2001, p. 267.
26. Baker, Lillian. *Plastic Jewelry of the Twentieth Century.* Paducah, Kentucky: Collector's Books, 2003, p. 29.
27. DuBois, J. Harry. *Plastics History U.S.A.* Boston, Massachusetts: Cahners Books, 1972, p. 289.
28. Lauer, Keith and Robinson, Julie. *Celluloid: Collector's Reference and Value Guide.* Paducah, Kentucky: Collector's Books, 2001, p. 20.
29. "Dupont Heritage: Plastics" Website, URL: http://www.heratige.dupont.com/touchpoints/tp1924/overview.shtml, 2003.
30. Interview with Steve Coulter, Technical Service Representative of Lucite International (Charterhouse Development Capital, Ltd.), 2003.
31. Interviews with Maureen Reitman, Senior Managing Engineer, of Exponent Inc., 2003, 2004.
32. DuBois, J. Harry. *Plastics History U.S.A.* Boston, Massachusetts: Cahners Books, 1972, pp. 290-292.
33. Interview with Norm Kaufman, former President of First in Imports, 2003.
34. DuBois, J. Harry. *Plastics History U.S.A.* Boston, Massachusetts: Cahners Books, 1972, p. 205.
35. "People & Polymers: Pierre Castan" Website, URL: http://Plastiquarian.com/castan.htm, 2004.
36. DiNoto, Andrea. *Art Plastic: Designed for Living.* New York, New York: Abbeville Press, 1984, p. 64.
37. Interviews with Maureen Reitman, Senior Managing Engineer, of Exponent Inc., 2003, 2004.
38. Ibid.
39. Interviews with Keith Lauer, Curator & Plastorian, National Plastics Center & Museum, 2003, 2004.
40. Ibid.
41. Lauer, Keith and Robinson, Julie. *Celluloid: Collector's Reference and Value Guide.* Paducah, Kentucky: Collector's Books, 2001, p. 170-171.
42. Interviews with Maureen Reitman, Senior Managing Engineer, of Exponent Inc., 2003, 2004.
43. Ibid.
44. Interviews with Plastics Engineers Rodney Rouleau & George Ivanov of Panduit Corp., 2003, 2004.
45. Ibid.
46. Interview with Carol Rowe of Plastics from the Past, Tucson, AZ, 2004.
47. Ibid.
48. Parry, Karima "About Testing", Website, URL: http://www.Plasticfantastic.com/testing.html, 2004.
49. Interview with Carol Rowe of Plastics from the Past, Tucson, AZ, 2004.
50. Interviews with Maureen Reitman, Senior Managing Engineer, of Exponent Inc., 2003, 2004.
51. Ibid.
52. Interviews with Plastics Engineers Rodney Rouleau & George Ivanov of Panduit Corp., 2003, 2004.
53. Ibid.
54. Interviews with Maureen Reitman, Senior Managing Engineer, of Exponent Inc., 2003, 2004. (URL:http://www.analyzeinc.com/web/contact.htm)
55. Ibid.
56. Interview with Bob Gluck, owner of Perfect Pearl, Skokie, Illinois, 2004.
57. Interview with Steve Coulter, Technical Service Repetitive of Lucite International, (Charterhouse Development Capital, Ltd.) 2003.
58. Interviews with Keith Lauer, Curator & Plastorian, National Plastics Center & Museum, 2003, 2004.
59. Battle, Dee and Lesser, Alayne. *The Best of Bakelite and Other Plastic Jewelry.* Atglen, Pennsylvania: Schiffer Publishing, Ltd., 1996, p. 7.

60. DiNoto, Andrea. *Art Plastic: Designed for Living.* New York, New York: Abbeville Press, 1984, pp. 43-44.
61. Interviews with Maureen Reitman, Senior Managing Engineer, of Exponent Inc., 2003, 2004.
62. Interviews with Keith Lauer, Curator & Plastorian, National Plastics Center & Museum, 2003, 2004.
63. Interviews with Maureen Reitman, Senior Managing Engineer, of Exponent Inc., 2003, 2004.
64. DuBois, J. Harry. *Plastics History U.S.A.* Boston, Massachusetts: Cahners Books, 1972, p. 218.
65. Kittler, Glenn D., *More Than Meets the Eye: The Foster Grant Story.* New York, New York: Coronet Books, 1972, pp. 81, 107.
66. Interview with H. Jack Feibelman, Former Director of Product Development for Coro, 2003.
67. Interview with Norm Kaufman, former President of First in Imports, 2003.
68. Interview with Burliegh Greenberg, former Executive Vice President and General Manager of Brier Manufacturing Company, 2003.
69. Interviews with Keith Lauer, Curator & Plastorian, National Plastics Center & Museum, 2003, 2004.
70. Kittler, Glenn D., *More Than Meets the Eye: The Foster Grant Story.* New York, New York: Coronet Books, 1972, pp. 1-2, 107.
71. Interview with Norm Kaufman, former President of First in Imports, 2003.
72. Ibid.
73. Interview with Burliegh Greenberg, former Executive Vice President and General Manager of Brier Manufacturing Company, 2003.
74. Interview with Bob Gluck, owner of Perfect Pearl, Skokie, Illinois, 2003.
75. "About Us: Company History", *Swarovski Company Website*, URL: http://www.swarovski.com, 2004.
76. Weisberg, Naida D., *Diamonds are Forever, but Rhinestones are for Everyone!* Providence, Rhode Island: The Providence Jewelry Museum, 1999, p. 18.
77. "The Precious Heritage of Trifari: Precious Heritage Backgrounder" from Trifari Company sales materials, 1981.
78. Interview with Karl Eisenberg owner of Eisenberg Ice, Chicago, Illinois, 2003, 2004.
79. Interviews with Iraida Garey, former Vice President of Product Development for Lisner, 2003, 2004.
80. Interview with Alan Marcher, former Salesman for Coro and Lisner, 2003.
81. Interview with H. Jack Feibelman, Former Director of Product Development for Coro, 2003.
82. Pamphlet "Jewels by Trifari Deceive Jewel Thieves" from Trifari Company sales materials, c. 1950.
83. Interview with Jim Axelrad, former owner of Pakula, 2004.
84. Interviews with Iraida Garey, former Vice President of Product Development for Lisner, 2003, 2004.
85. Interview with Larry Kassoff, former owner of Florenza, 2003.
86. "20401…by Trifari" from Trifari Company sales materials, c. late 1960s.
87. Ibid.
88. Ibid.
89. Ibid.
90. Pamphlet "Jewels by Trifari Deceive Jewel Thieves" from Trifari Company sales materials, c. 1950.
91. "10 Easy Questions…10 Easy Answers" from Trifari Company sales materials, c. mid-1970s.
92. Interview with Carole and Stan Smith, owners of Ralph Singer Jewelry, 2004.
93. "20401…by Trifari" from Trifari Company sales materials, c. late 1960s.
94. "Jewels by Trifari: Manufacturing" from Trifari Company sales materials, c. late 1970s.
95. Pamphlet "The Beauty of Trifari" from Trifari Company sales materials, c. 1980s.
96. Interview with Irving Wolf, former Chairman and Chief Executive Officer of Trifari, 2003.
97. Interview with Marsha Brenner, jewelry designer and former owner of Just Jewelry, 2003.
98. Interview with Carole and Stan Smith, owners of Ralph Singer Jewelry, Chicago, Illinois, 2004.
99. Interview with Jane Civins of the Providence Jewelry Museum, 2003.
100. Interview with Iraida Garey, former Vice President of Product Development for Lisner, 2003, 2004.
101. Interview with Marsha Brenner, jewelry designer and former owner of Just Jewelry, 2003.
102. Tooher, Nora Lockwood. "Hired Hands: The Women Behind the Rhode Island Jewelry Industry". *The Rhode Islander Magazine*, September 4, 1994.
103. Interview with Jim Axelrad, former owner of Pakula, 2004.
104. Tooher, Nora Lockwood. "Hired Hands: The Women Behind the Rhode Island Jewelry Industry". *The Rhode Islander Magazine*, September 4, 1994.
105. Jewelry Employment Data, Providence, Rhode Island, 1947-2004, Rhode Island Economic Development Corp., 2004.
106. Interview with Jim Axelrad, former owner of Pakula, 2004.
107. Interview with Karl Eisenberg owner of Eisenberg Ice, Chicago, Illinois, 2003, 2004.
108. Interview with H. Jack Feibelman, Former Director of Product Development for Coro, 2003.
109. Interview with Robert Mandle, former owner of the R. Mandle Company, 2003.
110. Interview with Iraida Garey, former Vice President of Product Development for Lisner, 2003, 2004.
111. Interview with Jim Axelrad, former owner of Pakula, 2004.
112. McCormick, Elsie, "Merchants of Glitter", *Saturday Evening Post*, May 31, 1947, pp. 40-41, 144-146.
113. Trifari, Krussman & Fishel, Inc. v. Charel Co., Inc., 134F.Supp.551(U.S.D. NY, 1955)
114. Interview with Jim Axelrad, former owner of Pakula, 2004.
115. Weisberg, Naida D., *Diamonds are Forever, but Rhinestones are for Everyone!* Providence, Rhode Island: The Providence Jewelry Museum, 1999, p. 85.
116. McCormick, Elsie, "Merchants of Glitter", *Saturday Evening Post*, May 31, 1947, pp. 40-41, 144-146.
117. Weisberg, Naida D., *Diamonds are Forever, but Rhinestones are for Everyone!* Providence, Rhode Island: The Providence Jewelry Museum, 1999, p. 20.
118. Interview with Carole and Stan Smith, owners of Ralph Singer Jewelry, Chicago, Illinois, 2004.
119. McCormick, Elsie, "Merchants of Glitter", *Saturday Evening Post*, May 31, 1947, pp. 40-41, 144-146.
120. Interview with Karl Eisenberg, Owner of Eisenberg Ice, Chicago, Illinois, 2003, 2004.
121. Weisberg, Naida D., *Diamonds are Forever, but Rhinestones are for Everyone!* Providence, Rhode Island: The Providence Jewelry Museum, 1999, pp. 53,143,162.
122. Interview with Irving Wolf, former Chairman and Chief Executive Officer of Trifari, 2003.
123. "Trifari Through the Years" from Trifari Company sales materials, 1991.
124. Interviews with Wendi Mancini, former Sales Representative for Trifari, 2003, 2004.
125. Transcript of Trifari Sales Representative Wendi Mancini's Lecture to Design Department Students, Western Illinois University, Macomb, Illinois, 1965.
126. Interview with Jim Axelrad, former owner of Pakula, 2004.
127. Interview with Marsha Brenner, jewelry designer and former owner of Just Jewelry, 2003.
128. Interview with Alan Marcher, former Sales Representative for Coro and Lisner, 2003.
129. Interview with Jim Axelrad, former owner of Pakula, 2004.
130. Interview with Alan Marcher, former Sales Representative for Coro and Lisner, 2003.
131. Interviews with Wendi Mancini, former Sales Representative for Trifari, 2003, 2004.
132. Interview with Geraldine King, former Fashion Jewelry Buyer for Marshall Field's, 2003.
133. Interview with Alan Marcher, former Sales Representative for Coro and Lisner, 2003.
134. Interview with Jim Axelrad, former owner of Pakula, 2004.
135. Interviews with Wendi Mancini, former Sales Representative for Trifari, 2003, 2004.
136. Interview with Jim Axelrad, former owner of Pakula, 2004.
137. Interviews with Wendi Mancini, former Sales Representative for Trifari, 2003, 2004.
138. Interview with Robert Mandle, former owner of the R. Mandle Company, 2003.
139. "Coro Company Records" Rhode Island Historical Society Manuscripts Division Website. URL: http://www.rihs.org/mssinv/Mss237.htm, 2003.
140. Brunialti, Carla Ginelli and Brunialti, Roberto. *A Tribute to America: Costume Jewelry 1935-1950.* Milan, Italy: EDITA, 2002, pp. 29-30.
141. "Coro Company Records" Rhode Island Historical Society Manuscripts Division Website. URL: http://www.rihs.org/mssinv/Mss237.htm, 2003.
142. "Costume Jewelry", *Fortune Magazine*, December 1946, p. 143.
143. Brunialti, Carla Ginelli and Brunialti, Roberto. *A Tribute to America: Costume Jewelry 1935-1950.* Milan, Italy: EDITA, 2002, pp. 29-30.

144. Ibid.
145. "Coro Company Records" Rhode Island Historical Society Manuscripts Division Website. URL: *http://www.rihs.org/mssinv/Mss237.htm*, 2003.
146. Interview with H. Jack Feibelman, Former Director of Product Development for Coro, 2003.
147. Brunialti, Carla Ginelli and Brunialti, Roberto. *A Tribute to America: Costume Jewelry 1935-1950.* Milan, Italy: EDITA, 2002, pp. 29-30.
148. Ibid.
149. "Coro Company Records" Rhode Island Historical Society Manuscripts Division Website. URL: *http://www.rihs.org/mssinv/Mss237.htm*, 2003.
150. Dolan, Maryanne. *Collecting Rhinestone & Colored Jewelry: 3rd Edition.* Florence, Alabama: Books Americana, 1993, p. 123.
151. Brunialti, Carla Ginelli and Brunialti, Roberto. *A Tribute to America: Costume Jewelry 1935-1950.* Milan, Italy: EDITA, 2002, pp. 29-30.
152. "Coro Company Records" Rhode Island Historical Society Manuscripts Division Website. URL: *http://www.rihs.org/mssinv/Mss237.htm*, 2003.
153. "High Beam eLibrary Results" of "Coro Richton" search from Women's Wear Daily Articles 1985-1992, High Beam Research™ Website. URL: *http://www.highbeam.com/library*, 2004.
154. Ibid.
155. Interview with Maureen Moloney, Market Researcher, Quebec Delegation of Chicago with Dunn & Bradstreet information provided by the Bureau du Commercial, Quebec, 2004.
156. "Eleven to Watch: Rainmaker" Forbes Website. URL: *http://www.forbes.ocm/forbes/2000/1030/6612194a_print.html*, 2003.
157. "How do they…Remain successful after decades of fashion" *Accent Magazine,* January 1991, pp. 51-53, 56.
158. Pamphlet "Jewels by Trifari Deceive Jewel Thieves" from Trifari Company sales materials, c. 1950.
159. Interviews with Wendi Mancini, former Sales Representative for Trifari, 2003, 2004.
160. "The Precious Heritage of Trifari: Precious Heritage Backgrounder" from Trifari Company sales materials, 1981.
161. "Costume Jewelry", *Fortune Magazine,* December 1946, p. 145.
162. "The Precious Heritage of Trifari: Precious Heritage Backgrounder" from Trifari Company sales materials, 1981.
163. Interview with Irving Wolf, former President of Trifari, 2003.
164. Cohen, Judith W. and Weil, Jeanne. "Jews in the Jewelry Industry in Rhode Island, Part I" *Rhode Island Jewish Historical Notes,* Vol.10, No. 3, Part B, November 1989, p. 287.
165. Interview with Irving Wolf, former President of Trifari, 2003.
166. Brunialti, Carla Ginelli and Brunialti, Roberto. *A Tribute to America: Costume Jewelry 1935-1950.* Milan, Italy: EDITA, 2002, p 42.
167. "Our Brands" Liz Claiborne Inc. Website, URL: *http://www.lizclaiborneinc.com/ourbrands/default.asp*, 2003.
168. Rainwater, Dorothy T., *American Jewelry Manufacturers.* Atglen, Pennsylvania: Schiffer Publishing Ltd., 1988, p. 157.
169. Interview with Tony Ganz, son of Victor Ganz, 2003.
170. Brown Marcia and Tempesta, Lucille, "The Magic of Mandle", *Vintage & Fashion Costume Jewelry Magazine,* Vol. 13, No. 1, Winter 2003, pp. 8-13.
171. Dolan, Maryanne. *Collecting Rhinestone & Colored Jewelry: 3rd Edition.* Florence, Alabama: Books Americana, 1993, p. 102.
172. FitzGerald, Michael. *A Life of Collecting: Victor and Sally Ganz.* New York, New York, Christie's, Inc., 1997, pp. 8-24.
173. Interviews with Iraida Garey, former Vice President of Product Development for Lisner, 2003, 2004.
174. FitzGerald, Michael. *A Life of Collecting: Victor and Sally Ganz.* New York, New York, Christie's, Inc., 1997, p. 219.
175. "The New York Times" On-Line Picasso Project Archives Website excerpt from New York Times article "Prized Picasso Leads Collection to Record $206-Million Auction" by Carol Vogel, November 11, 1997, URL: *http://www.tamu.edu/mocl/picasso/archives/opparch97-188.html*, 2003.
176. Interview with Kate Ganz Belin, daughter of Victor Ganz, 2003.
177. Interview with Vicky de Felice, daughter of Victor Ganz, 2003.
178. Interview with Kate Ganz Belin, daughter of Victor Ganz, 2003.
179. Interviews with Iraida Garey, former Vice President of Product Development for Lisner, 2003, 2004.
180. Interview with Kate Ganz Belin, daughter of Victor Ganz, 2003.
181. Rainwater, Dorothy T., *American Jewelry Manufacturers.* Atglen, Pennsylvania: Schiffer Publishing Ltd., 1988, p. 170.
182. Interview with Robert Andreoli former owner of Lisner-Richelieu & Victoria Creations, 2004.
183. Weisberg, Naida D., *Diamonds are Forever, but Rhinestones are for Everyone!* Providence, Rhode Island: The Providence Jewelry Museum, 1999, pp. 33-34.
184. "Company History" Jones Apparel Group Website, URL: *http://ir.thomsonfn.com/InvestorRelations/PubBusinessOutlook.aspx?partner=9189*, 2004.
185. Interview with Robert Andreoli, former owner of Lisner-Richelieu & Victoria Creations, 2004.
186. Brown, Marcia. *Unsigned Beauties of Costume Jewelry.* Paducah, Kentucky: Collector's Books, 2000, p. 201.
187. Interview with H. Jack Feibelman, Former Director of Product Development for Coro, 2003.
188. Interview with Robert Mandle, former owner of the R. Mandle Company, 2003.
189. Brown, Marcia. *Unsigned Beauties of Costume Jewelry.* Paducah, Kentucky: Collector's Books, 2000, p. 201.
190. Interviews with Iraida Garey, former Vice President of Product Development for Lisner, 2003, 2004.
191. Romero, Christie, *Warman's Jewelry Identification and Price Guide.* Iola, Wisconsin: Krause Publications, 2002, p. 254.
192. Interview with H. Jack Feibelman, Former Director of Product Development for Coro, 2003.
193. Interview with Larry Kassoff, former owner of Florenza, 2003.
194. Interview with Cheryl Killmer, owner of Past Perfection, URL:http://www.pastperfection.com, 2004.
195. Interviews with Maureen Reitman, Senior Managing Engineer, of Exponent Inc., 2003, 2004.
196. "Signs of Deterioration" Plastics Historical Society Website, URL: *http://www.plastiquarian.com/deterio.htm*, 2003.
197. Ibid.
198. Lauer, Keith and Robinson, Julie. *Celluloid: Collector's Reference and Value Guide.* Paducah, Kentucky: Collector's Books, 2001, pp. 258-264.
199. Interview with Cheryl Killmer, owner of Past Perfection, URL:http://www.pastperfection.com, 2004.
200. Vintage Fashion & Costume Jewelry (VFCJ), P.O. Box 265, Glen Oaks, NY 11004, e-mail: VFCJ@aol.com.

Selected Bibliography

Baker, Lillian. *Plastic Jewelry of the Twentieth Century*. Paducah, Kentucky: Collector's Books, 2003.

Battle, Dee and Lesser, Alayne. *The Best of Bakelite and Other Plastic Jewelry*. Atglen, Pennsylvania: Schiffer Publishing, Ltd., 1996.

Brown, Marcia. *Unsigned Beauties of Costume Jewelry*. Paducah, Kentucky: Collector's Books, 2000.

_____. *Signed Beauties of Costume Jewelry*. Paducah, Kentucky: Collector's Books, 2002.

Brunialti, Carla Ginelli and Brunialti, Roberto. *A Tribute to America: Costume Jewelry 1935-1950*. Milan, Italy: EDITA, 2002.

DiNoto, Andrea. *Art Plastic: Designed for Living*. New York, New York: Abbeville Press, 1984.

Dolan, Maryanne. *Collecting Rhinestone & Colored Jewelry: 3rd Edition*. Florence, Alabama: Books Americana, 1993.

DuBois, J. Harry. *Plastics History U.S.A.* Boston, Massachusetts: Cahners Books, 1972.

Kelley, Lyngerda and Schiffer, Nancy. *Plastic Jewelry*. Atglen, Pennsylvania: Schiffer Publishing Ltd., 1987.

FitzGerald, Michael. *A Life of Collecting: Victor and Sally Ganz*. New York, New York, Christie's, Inc., 1997.

Katz, Sylvia. *Plastics: Commonplace Objects, Classic Designs*. New York, New York, Harry N. Abrams, Inc., 1984.

Katz, Sylvia. *Early Plastics*. Buckinghamshire, U.K., Shire Publications, 1994.

Kittler, Glenn D., *More Than Meets the Eye: The Foster Grant Story*. New York, New York: Coronet Books, 1972.

Lauer, Keith and Robinson, Julie. *Celluloid: Collector's Reference and Value Guide*. Paducah, Kentucky: Collector's Books, 2001.

Miller, Judith. *DK Collector's Guides: Costume Jewelry*. New York, New York: DK Publishing, Inc., 2003.

Rainwater, Dorothy T., *American Jewelry Manufacturers*. Atglen, Pennsylvania: Schiffer Publishing Ltd., 1988.

Robins, Natalie and Aronson, Steven M.L., *Savage Grace*. New York, New York: Dell Publishing Co., Inc., 1985.

Romero, Christie, *Warman's Jewelry Identification and Price Guide*. Iola, Wisconsin: Krause Publications, 2002.

Weisberg, Alfred M., *Why Providence?* Providence, Rhode Island: Technic, Inc., 1992.

Weisberg, Naida D., *Diamonds are Forever, but Rhinestones are for Everyone!* Providence, Rhode Island: The Providence Jewelry Museum, 1999.

Index